TIME TRACKS
SCENES FROM THE IRISH EVERYDAY

D1794721

TIME TRACKS

SCENES FROM THE IRISH EVERYDAY

MICHAEL CRONIN

NEW
ISLAND

Copyright © 2003 MICHAEL CRONIN

TIME TRACKS
First published 2003
by New Island
2 Brookside
Dundrum Road
Dublin 14
www.newisland.ie

The author has asserted his moral rights.

ISBN 1 904301 41 X

British Library Cataloguing in Publication Data. CIP catalogue record for this book is
available from the British Library.

Typeset by New Island. Cover design by New Island. Printed by CPD, Ebbw Vale, Wales.

New Island received financial assistance from
The Arts Council (An Chomhairle Ealaíon), Dublin, Ireland.

10 9 8 7 6 5 4 3 2 1

For Juliette

ACKNOWLEDGEMENTS

Acknowledgements are due to the editors of *The Dublin Review* and the "Writing Now" section of *The Irish Times* where two of the pieces in this collection first appeared.

CONTENTS

RICH TEAS

A LUNCH IS BEING HELD IN HONOUR OF A FRENCH writer who is in Dublin to organise a writers' festival. The conversation turns inevitably to food and, as often happens with a large group around a table, the general conversation breaks down into sub-plots of chat. After a while, the visiting writer notices the animation of your neighbours at the table and wants to know what is the reason for all this excitement. He seems puzzled when you tell him – biscuits.

But he need not have been so surprised. Marcel Proust unlocks his past not with a rare steak or a slice of pâté de foie gras but with a small teacake, a madeleine. Sweet dreams are our key to memory. The naming of biscuits was for the Irish guests at the table a guided tour through childhood – the sugared crumbs, a trail of recollection. Marietta was there at the beginning of the path. The round coin of austerity with its suggestion of poorly heated parish halls and convent parlours and prudent excess. Or else the dismembered Marietta biscuits reduced to the scale of the doll's world, sustaining Sindy or Barbie through those endless afternoon teas of the winter months, miniature dramas played out with droplets of milk in scratched red plastic cups and the occasional scolding from the director when Barbie fouled her twin set with great boulders of biscuit. Butter redeemed Marietta's puritan plainness. Spread on two biscuits that were then put together, the butter that oozed through the pinpricks on the surface like inquisitive earthworms carried with it the promise of luxury and a faint intimation of decadence.

Digestive biscuits in their gritty wholesomeness were associated like all things that are good for you with the sickroom. In the same category as coarser breakfast cereals, they suggested disciplined recovery from mumps or the measles or a bad flu. Hot lemon drinks and the tentative crumbling of the plain digestive biscuit broke up the sick-day routine of endlessly reread *Treasure* and *Tiger and Jag* comics from the last school garden fête – where it always rained despite the promise of Gallic summer in *fête*. Chocolate on digestive biscuits was always confusing, like a category confusion in logic, so that these biscuits were consumed with something like lingering guilt, as if you were found stuffing yourself with Belgian chocolates on a health farm.

Rich Tea and Morning Coffee. The decorous abandon of late morning, the fountain trickle of talk radio and the house huge with the silence of the children gone to school or a day snatched from work, yawning with possibility. The biscuits that adorn the dark tables of meetings never seem to have the same effect, as if the suggestion of leisure is scotched by the seriousness of purpose, the tea going cold and the biscuits, more worry beads than sweetmeats, dismembered by fidgety fingers. Afternoon Tea Assorted. The panic of want as the unexpected arrival of a neighbour or a relative (*Just thought I'd say hello, Margaret*) means turmoil in the kitchen. The only biscuits left are four Lincolns, soft with age, and shards of cream cracker. Behind the welcoming smiles are urgent, whispered commands to go to Donnelly's and buy a packet of biscuits. While your bike carries you to the shop, your mother is back at the house talking to the neighbour-relative as if she was an accomplice in a robbery, distracting the police officer's attention with bogus requests for directions to go to Blackhall Place or Constitution Hill. As the tray arrives with the tea and the biscuits are spread in a bountiful landslide on the plate (not too large, of course, otherwise the biscuits thin into insignificance), the complimentary guest (*How big you are now! What class are you in?*) secretly appraises the offering. Excessive plainness in the biscuit is a subtle affront, the Marietta snub. Chocolate fingers or Bourbons suggest a social ambitiousness that needs to be watched. Away from the grand manoeuvres of the Good Room, there is, however, the campfire of gossip in the warm kitchen where biscuits are dunked and time dissolves in the Great Plains of a weekday afternoon and the doilys stay in the sideboard.

With Empire came the lure of the exotic. Kimberley, Mikado, Coconut Creams. A late Victorian map of plenty, handlebar moustaches and steamships plotting passages for the cargo of raw materials and spices from the four corners of the earth. But our minds soon tired of the *Look and Learn* immensity of overseas possessions and began to focus in on the one abiding passion of childhood – demolition. The trick was to use your front teeth like a barber's razor, removing the two white banks of coconut, leaving the thin stream of strawberry in the middle, resting on its biscuit base. A similar gopher-like application was used to remove the coconut mound in another biscuit from its base: the cleaner the cut, the keener the triumph. Chocolate Snack biscuits were other contenders for the demolition derby. Here, the aim was to remove chocolate in a series of quick, decisive bites until only the biscuit was left, pale, slightly moist, the cream fragment of bone exposed by the energetic dig. The pleasure was in removing whole sides of chocolate in one carefully aimed nip. If only a piece came away, or the top section had to be scraped off, there was a vague sense of irritation, an irritation that returns in adult life when taking off wallpaper, obstinate islands of ancient paste interrupting the triumphant march of the paper stripper. Chocolate mallows involved more delicate surgery. Here, the fine chocolate membrane had to be removed from the sticky dome underneath while leaving the dome intact. Only when the marshmallow lay exposed, vividly white, could the next operation begin: the severing of the marshmallow from its base. As the sweet oyster dissolved in your mouth, the final act was to eat the base itself, moist now and striated with the ploughed ridges of teeth marks.

Biscuits were our Gold Standard. All important events in early life were quantified by their sweetness. As the coins and notes were slipped furtively into your First Holy Communion palm, the cogs of the great machines of imperial division turned in the counting-house of the mind. Two pounds, fourteen shillings and sixpence. Twelve pennies in a shilling, twenty shillings in a pound. The lava spill of coins on the bedspread and the uneven columns of halfpennies (there were some, from the witch in number ten), thruppences, sixpences, shillings, half-crowns, rising from the bedside locker. The orange and green notes, exotic rugs in this shattered Parthenon, refusing to stay flat, with blue-biroed telephone numbers scribbled in their grubby margins. A man has two pounds, fourteen shillings and sixpence. He goes to a shop. The biscuits cost threepence each. How many biscuits will he get for his money? Levers were pulled and cogs meshed as the lordly denominations of half-crowns and shillings were reduced to the workaday, proletarian brownness of the penny. Two pounds, fourteen shillings and sixpence: that makes (frown, pursed lips) six hundred and fifty-four pennies. The noise of the machine is now a roar as the mind runs from dial to dial, excited and alarmed at the monstrous prospect before it. Six hundred and fifty-four divided by three (find a pencil and tear open an empty Senior Service box): let's see, that makes, two hundred and eighteen. Two hundred and eighteen Disc-O-Chocs! The bodhrán-rattle of the heart increases as you repeat the arithmetical operations and, a dizzy Edisonian wunderkind, you stagger through the failing light of the basement workshop, numbed by your discovery. Two hundred and eighteen Disc-O-Chocs!

This is storybook wealth, a cave of sweetshop legend. In your short pants and communion blazer, you are an unlikely candidate for the grim contentment of advanced age, pottering around in Mutual Growth Funds and indexing happiness against the FT, NASDAQ and Dow-Jones. But now riches beyond reckoning lie waiting in Mrs Byrne's shop – the chocolate biscuits individually wrapped in heavy foil, piled up like so many florins in the pirate coffers of the glass-fronted counter. The expression on Mrs Byrne's face as you demand two hundred and eighteen Disc-O-Chocs. The look of momentary panic and then the whispered instructions to Sheila (the one with the Red Hurley badge) to go to the storeroom; the bank-robbery tenseness of the scene as the boxes of Disco-O-Chocs are brought in nervous haste from the vaults and you back out of the shop, sweaty palm strangling the top of the transparent sack, a whole lifetime of idle nibbling in store for the Disco-O-Choc Kid. You never manage to carry out the raid. Long before you turn up at the premises, most of your money has been impounded by the Juvenile Assets Bureau (JAB) in the form of your parents who hand you back a light-green Post Office Savings book in return for the bedside hoard. When you do manage to buy a Disc-O-Choc with a salvaged thruppence, you stand outside Mrs Byrne's, a forlorn refugee from a speculative crash, your money now gone in some magical abstraction of high finance, biting into the chocolate surface and resigning yourself to a life of honest toil and Arrowroot.

In geography class, national wealth is again a matter of plates piled high. When we progress around Ireland on the Stations of Memory, stopping in each town and city to repeat the principal economic activities of these urban centres, the great plainchant of enumeration filling the classroom – brewing, boot-making, coal-mining (Castlecomer), textiles, sugar (Clonmel) – very occasionally in the grey fog of industry, the word "biscuits" would stand out. Along with stout, whiskey, close relatives and bellowing cattle, it was one of our few export industries. Bolands and Jacobs were the Gog and Magog of this world, two

giants looming over the shelves, fixing their
followers with a fierce, possessive stare. There were
Jacobs' families and Bolands' families, just as in
France in later years you could tell what people
read for breakfast and how they voted once you
knew what car they drove (Citroën or Renault).
Then there were the double agents, who went with
Bolands for the Cream Crackers but stayed with
Jacobs for the Bourbons. A friend in school had an
aunt who worked in the Jacobs' factory in Tallaght
and he seemed possessed of a special grace, an
infant lama, the cover of his Tupperware lunch-box
peeled away to reveal the sacred treasure of
Coconut Creams, Lemon Puffs and Club Milks.
Like mendicants scurrying after the sahib's taxi, we
courted him and flattered his interest in racing cars
(*Yeah! Jackie Stewart! Yeah!*) in the hope that some
of his family fortune might make its way into our
own lunch-boxes. He had, however, all the
canniness of a Renaissance prelate and though we
heard many tales about the opulence of the court of
W. and R. Jacob, we remained obstinately outside,
cursing our parents' career choices and fingering
with disdain the Calvita-filled triangles of white
bread and the greasy surface of the Granny Smith.

In the dull tramp of necessity down the supermarket aisles, loading up milk and butter and rice
and sausages, only the biscuit shelves seemed to
offer a holiday of choice. Being allowed to choose
the biscuits by your mother was an obscure but
momentous favour, soured only by the exasperated

grimace from the end of the aisle that We Hadn't All Day. Not total freedom of course. Nations still guarded our larders so the sight of McVities or Burtons in the trolley meant the swift rebuke of dead generations and the immediate suspension of all privileges. Anything with chocolate was expensive so you had to instinctively balance the budget of desire with the harsh realities of till receipts. Later, as the trolley is replaced by the flatland basket and queasy Saturday morning expeditions to the mini-market, your mouth musty from last night's fun, there is the crossing of some invisible threshold as you surrender the right to choose the biscuits for your bedsit to your girlfriend. These increments of trust do not make for the high drama of romantic prose (*He turned ever so slowly and said to me, "Take them, darling, I'm willing to try them out"*) but they do map out their own intimacies. Sometimes, of course, nothing works out and you are left mournfully masticating her choice – rectangular Nice biscuits that you never liked anyway.

One day you drop by to see your friend from college and he circles the kitchen, a haunted man. He talks but he is not listening and every so often his gaze travels to the white Formica cupboard with the scallop-ended silver handles. A jittery Raskolnikov, fiddling with guilt and wondering when the unmarked car will pull into the car-park opposite. No, not quite, but he is being picked at by conscience. Finally, the pints in the late evening lounge draw the confession. His Italian girlfriend has issued a formal edict, no more biscuits in the flat. Eating between meals is bad for you. Rots your teeth. Like animals in a sty, forever grunting and snuffling and licking the trough of early morning, late afternoon, midnight snacks. Look at how big they are in the United States. No appetite left for their dinner. The sacrifice seemed minimal in return for the brown-eyed attentiveness of Alessandra. He dutifully ate his luncheon apple and sucked in his dinnertime banana, smiling at his virtue and Alessandra's blue toenails. He even felt lighter, his teeth less clogged, more pence in his pocket. Like an ex-smoker in the giddy first week, he marvelled at the apparent ease of this transition to a world without sweetness. In the local Spar he headed straight for dairy products, his eyes chastely averted from the come hither wantonness of Raspberry and Custard Creams ranged on both sides of the aisle. Like a busy politician, he politely declined the fig rolls on a side plate offered to members at his local Amnesty branch meeting.

He alarmed his mother by leaving a saucerful of Jaffa Cakes untouched, the biscuits an obvious bribe to get him to come out to see her more often (as she polished off the cakes later on that evening she wondered what she had done to deserve him). But he was finding the absoluteness of his vows increasingly intolerable. He felt every billboard, every magazine, every television advertisement (poured chocolate was the most poignant reminder of the erotic motives of his own sacrifice) was taunting him in his Lenten solitude. The darkest night was mid-morning. The pot of tea and the newspaper in the empty flat, long a blessed island of meditative peace, were now dauntingly in-complete. A basic element of the ritual was missing so the Gods of ease stayed away, refusing to be summoned by the hand groping blindly from behind the newspaper for biscuits that were no longer there. Even Alessandra's vigorous afternoon lovemaking lacked the essential afterword of dazed conversation over Ginger Nut and Lyons Green Label. Your friend began to thrash about for the alibis of relapse. His sugar levels were too low. Maybe he was a diabetic but he had never known before. Young men need lots of energy (for sports he never played, he admitted in his more lucid moments). Before long, he was back on the habit again. His life now involved furtive expeditions to the supermarket on the other side of the river, adulterous circumspection in the incineration of empty packaging and the disposal of crumbs and the careful removal of chocolate smears from the corners of his lips. As the call for last orders rocked the intercom, your friend said he could no longer continue with this double life, he must make a clean breast of it (his words, not mine) and he was not going to give up all that Alessandra meant to him for a packet of Blackberry Rolls.

You met the friend two months later outside the twenty-four-hour shop in Westmoreland

Street. The girl he was with had blue eyes. In her left hand she was holding his pale, pink paw and in her right a Jacobs Coconut Cream Gang Pack.

The bare necessities of life are a matter of taste. Dieticians are routinely appalled by the dental carelessness of the poor, tut-tutting the sugar sandwiches and the jam doughnuts and the endless tuck-feasts of biscuits. Glossy leaflets detail their low nutritional value and make appalled comparisons with storybook fruit and vegetables. Grim budgets highlight the debt trap of the sweet tooth. Like televisions in prison cells, biscuits on the tables of the destitute are an affront to the right-thinking (*Isn't well for them to be sitting there eating their chocolate Polos – more than you'd ever see in this house, I can tell you*). But poverty is a long sentence. No matter how many competitions you enter, no matter how many times you write in not more than ten words, "The reason I like not being poor is", you are condemned to the endless trench-days of private miseries. So when you rummage under the cushions of the sofa or rifle the pockets of old coats in the wardrobe for the forgotten manna of small change, your thoughts are not on the dizzying vitamin count of lentil soup but on the lotus moments of tea and a full packet of biscuits. The vanilla-scented, cream-filled squares eased with the gentle pressure of two fingers from a pack, listing heavily on the chipped kitchen table are the sacred hosts of forgetfulness, a brief supper, a snatched cigarette in the no man's land of Final Notices and school bills.

Christmas took the biscuit. The tins piled precariously high in the local supermarket were bright advertisements of abundance. The memory would linger through the year as the tins were recycled as resting places for shoe polish and J-Cloths, the incongruous clutter replacing the orderly maze of the red paper partitions in the newly opened box. The Empires of the Boxes were from the New World. USA Assorted, a ritzy invitation to the melting pot. All eyes focused on the round chocolate biscuit with the jellyfish spread languidly over the hole in the middle. With the swiftness of Fagan's pickpockets, small hands snatched this piece of biscuit exotica and the tongue savoured the dissolving sugar of the jellyfish, wrested by beaver teeth from the chocolate base. One night in Barcelona, the Catalan wife of an Irish friend told me that when she thought of Ireland, two words came to mind – tea, biscuits. Tea and biscuits, this indeed is where the Remembrance of Things Past might begin, in the pockets of detail, the crumbs from our tables, the touchstones of memory.

THE CHANGING
OF THE CUPS

MISTER NOLAN, A SMALL, BALDING MAN ON A HIGH stool, his foot swinging like a hanging metronome. He told us the story. A story of islanders who scanned the water for wreckage. "The good fortune of misfortune," he would whisper softly. The boxes floating on the water, bobbing into land. The stencilled code not giving much away. And then the crack, the fracture and the contents spilling into uncomprehending hands. What in the holy …

Later, the secret is out. The parish priest sits down to a feed. The fish he knows, and the burst potatoes, but there is no name for the other green vegetable piled high on a corner of the best plate in the house. *A surprise, Father. A surprise.* A nervous rabbit, he nibbles the tip of the forkful of weed before his mouth. The bitter taste puzzles him. But the juice, the juice is a give-away. He is eating tea-leaves.

Imagine that! Eating tea! Wasn't that the quare

one! Mister Nolan clipped his knee with his right hand and laughed and laughed. This was our first taste of anthropology. We tried to imagine a people who did not know what tea was but the effort was too much. Mister Nolan must be making it up. Nobody could be that stupid.

Tea cosies. Your sisters were always knitting them. Like the scarves that curled, they were rarely finished, discarded remnants of schoolroom exercises, padding the bottom of overstuffed drawers. Covering the brown earthenware teapot of your aunt, the tea cosy had a special significance, a mark of homely distinction, of a piece with the moist teacake and the crumbly scones. There was one with a pom-pom on the top of it, and you wondered how your aunt got the water in. You vaguely yearned for the reassuring cosiness of the cosy, something to camouflage the scratched aluminium of the teapot at home with its broken Bakelite handle. But they remained suspended on knitting needles or coiled in punctured plastic bags, a clutch of assorted woollen threads that never took shape on the table. Later, serving tea in a pot with a tea cosy seemed tragically dated or slightly embarrassing, like been found wearing a bodice with cream-coloured buttons or holding up your jeans with a *crios* belt.

Tea bags. Once square, now round. What do you do with tea bags? Is that what saucers are for? The place where you put the tea bag, that unsightly, dark-green sandbag listing heavily against the cup, slumped in a suspect pool of brackish water? They were an end to clutter, sealed containers of orderliness. The memory of the sink in the kitchen: the tea-leaves collecting around the plug-hole like an eternal, rain-swept autumn, sodden leaves blocking the drains. The good fortune of the leaves: your mother remembered being told that tea-leaves floating on the top of the tea meant that presents were on their way. The time-frame of course was never specified so despite the fact that the prophecy was always eventually fulfilled (Christmas, birthdays) you could never be quite sure if the leaves were just leaves or intimations of unknown riches. The more arcane lore of picking out the future in the arrangements of leaves at the bottom of a drained cup was unknown to us and left to experts in caravans and special-feature writers, short of a story. As the Age of Reason brought you from short pants to scepticism, you were less convinced by the hidden language of the floating world of the teacup, sensing the Christmas hoax in another guise. But for years you never liked tea strainers. Because you never know, do you. There just might be a connection. And who wants to scoop up those tiny fragments of hope before they even get a chance to drift enticingly on the tea's milky surface? It was tea bags not strainers which in the end did for the Diviners. No more presents. No more teatime teasing of the leaves – the child's spoon gently pushing the reluctant vessels out on to a pale, grey lake. No more lingering expectation that the leaf might, just might, come wrapped in shining, striped paper with the delicious rip-cords of ribbons. Now, the bag floats sullenly on the surface like some tragic victim of a drowning or is rescued, all dripping, to shrivel up, an ashen reminder of our diminished present.

Coffee or tea? Before, the question was rarely asked, coffee a remote phenomenon. In cafés, people never drank *café* but tea. Then the screw-top jars and the heaped teaspoonfuls of brown sand flicked into the blue-striped mugs. Americans drank coffee and we drank tea. The nettle of ambition in the coffee cup. The foreign holidays brought back complaints that you-could-never-get-a-decent-espresso in Dublin. And then as the country leaned into modernity, coffee became the talisman of progress. Dublin's Café Society. Cork's Cappuccinos. Galway's Gold Blend. Coffee or tea? As if you were being asked to choose between the turf banks and the Left Bank, between the frugal piety of the linoleum kitchen and the summer-tabled elegance of the boulevard. Coffee or tea? The shock troops of progress pounding around the perimeter fences of the Celtic Zoo always drank coffee. Strong. Black. Concentrated. I am not human. Until I get my fix. The producers sauntered smiling through the brightly coloured advertisements of Latin America but the consumers raced against the caffeine clock, strained wrist pawing the mouse and muttering mouth gargling sweet, cold coffee. The pots and coffee machines hissed and spat in the houses and apartments of the newly mobile like model-railway steam engines heralding a new dawn of industry. Tea drinking becomes associated, like holy-water fonts in the hallway and antimacassars on the sofa, with an earlier age of piety, propriety and poverty. Even charities organise "coffee mornings" as if good works need the alibi of the stimulant, a "tea morning" suggesting Oriental permissiveness and the infinite extension of the tea break, the bugbear of the Outraged Listener (*Them Corporation fellas! Sitting on their behinds all day, guzzling tea, and we paying for it!*).

Eros had preferences. You were never invited "back" or "up" for tea, it was always coffee. As if the difference between mothers and daughters was that the mothers offered you tea and the daughters coffee. The long moments of distracted chat as your hands warmed themselves around the mug and you gradually sobered up, wondering were there any biscuits in the flat.

The bare necessities of the bed-sit. The spoon scraping away shavings of sugar from the sides of a pink-coloured sugar bowl, cocaine crystals of sweetness salvaged from want. The eternal Sunday morning with the dog-eared carton of milk on the table. The nose sniffing anxiously around the jagged opening for sourness. Tea without milk was a penitential exercise, the detail that impressed you above all others in uncles' descriptions of trips to Lough Derg. You had a cousin who stayed a night on her way back from Germany. With your sisters you watched in appalled silence as she drank black tea in your mother's house, incontrovertible proof that abroad was a strange place and that beyond Dun Laoghaire was a World turned Upside Down. Later when you go to stay with friends in France, alarm takes hold of you when you discover that there is no such thing as a kettle in the apartment in Orléans. All of a sudden, tea-making becomes an exotic, alchemical exercise as your pale, freckled face scrutinises the water in a saucepan that takes forever to come to the boil. There is no teapot either, so you spoon the leaves into the swirling contents of the pan and spill half the contents of this strange tea-soup on to the table as you tilt the cauldron into a coffee cup. Later, in continental cafés, there was the inevitable Liptons' tea bag with the yellow label drooping from a white teapot. But no milk. Only a slice of lemon. So that even on the boulevard you could hear the whispered rosaries of Patrick's purgatory.

The simple blocs of this Cold War world. Lyons Green Label and PG Tips. Nobody ever knew what the PG stood for and no one could ever pronounce the "T" in "Tips" (coming hard on the "G") so everybody said PG "Chips". Things became more complicated when you realized that in the Deep South they drank Barry's Tea. For years, drinking a tea that nobody else around you drank, rather than accent or a superior hurling team, seemed incontrovertible proof of Cork separatism. When you were invited to a different house or stopped off somewhere to have tea and a "hang" sandwich, there was always the moment of barely concealed apprehension as your eyes checked the complexion of the tea's surface (the bone whiteness of watery tea, a sign of in-experience – friends – or tightness – enemies) and your tongue carefully probed the brew for quality of water (never the same abroad), strength of tea-leaf (Oh God! The heart staggers, that Yellow Label stuff!) and freshness of milk (the suspicions of the nose confirmed by the acrid taste of milk that had gone off and the imminent confrontation with your parents over This Stuff that You Couldn't Possibly Drink). There were safe houses for tea drinking and others where the barely moistened lips of the visitors told of social agonies and an eternity of lame excuses. Not being able to produce proper tea could only be excused on the grounds of ill health (the strange herbal teas of the incapacitated) or having lived outside the State during the formative years of tea making.

On the way home from school, the huts of corporation workers were the corrugated iron temples of the Irish Tea Ceremony. The instruments of the cult – the crumpled, white bag of sugar covered in a delta of brown streaks, the half-full milk bottle, the blackened kettle, the dented teapot and the chipped mugs. The tea break seemed an improbable luxury in the cement-whitened baldness of the new city, the workers strangely privileged in this al fresco ritual of tea drinking. The watery egalitarianism of tea. The great leveller. The parish priests and the revolutionaries, the teetotallers and the topers, men and women, all united by the imperial gift of the dark leaves. The day you go to meet the President with your fellow translators, after the mumbled words and the stricken smiles in front of instamatics, the teapots line up to be emptied. The President apologetically explains that she is the only one to be served on a small silver salver, as if this aristocratic privilege in the Vice-Regal Lodge is out of keeping with the tea-drinking collectivism of the young Republic.

The cold tea of the interview.

The stewed tea of canteens.

The fresh pot of crisis (*Sit down there, John, and I'll put the kettle on*).

The tea that tastes differently coming out of a flask on a picnic or being poured into a plastic cup on a school desk. The slightly incongruous sight of the steam rising from the cup in the frozen wastes of the prefabricated classroom. But the taste was always vaguely metallic, as if even the most thorough rinsing of the hall of mirrors inside the vacuum flask could not wash away the lingering odour of the earlier gallons of tea that had settled in the shrunken fun fair. The flasks were fragile things, of course, and schoolbags were not, so the tea breaks were usually cut short by the horrified realisation that the bag swung in anger had just led to the shattering of the tartan-patterned vessel, the warm cylinder now a useless kaleidoscope of silver shards.

Inside the mahogany sideboard, the great mausoleum of social ambition in the front room, they lay unused for months on end. The Good Cups. White bone china with gold rims and a picture of yellow roses. A wedding present that had survived four children to remain more or less intact. Breaking one of the Good Cups was a capital offence and the only hope for the offender was that the crime would go unnoticed. The arrival of a member of the clergy or the medical profession would involve the immediate summons to Bring Out the Good Cups. Entering the room with its eternally drawn curtains and unsoiled wallpaper was like a descent into the vaults of a ruined church, the spectral chill of unused space adding to the solemnity of the moment. Opening the door of the sideboard confirmed that the cups and plates and saucers were still there. The next difficult decision was how many pieces of crockery you could manage at any one time. The saucers and plates were a cinch but the cups were tricky. You might manage three on their side but you could only get two upright. Three upright meant one would topple over (pure speculation on your part: the consequences of being proved right through empirical experiment were too awful to contemplate). The thin, brittle edges of the cup made you nervous when you poured, as if the tea would crack the pale skin of fired clay. The cups themselves were not held like other cups and even the adults looked more like marionettes as they cautiously replaced the cup on the saucer and or held the cup stiffly in mid-air, frozen in a distant comedy of manners from the colonies played out in the sitting room of a suburban house. As the years roll into each other, the Edwardian apprenticeship of teatime gives way to the woolly jumper Communion of the Mug. The first sight of mugs was on American sitcoms and you wondered at first how these clumsy, ugly things could ever evict teacups from the tables of your contemporaries. But the mug was Youth. More evidence of the generation gap. While your father looked distractedly around for the absent saucer, you clung triumphantly to the mug, scorning the bourgeois propriety of the saucer and seeing the lumpish mug as a sign of a new egalitarianism. Mugs were *de rigueur* at the meetings of the Solidarity Committees and the Editorial Collectives and the Action Groups. Producing cups and saucers would have been tantamount to class betrayal or evidence of incorrigible young fogeyism, a dangerous lapse back into the cosseted pieties of another age. So you put up with the circular stains on the tables, a small cost for the campfire comradeship of the mug. A problem still remained – what to do with the tea bag?

With age comes choice. The first time you are invited to tea rooms in a fashionable hotel and all of a sudden there is not Tea but Teas. The bewildering maze of plurality. Earl Grey. Lapsang Souchong. Orange Pekoe. Blackcurrant. Peppermint. Raspberry.

Herbs and fruits piled up on the table of this Mad Hatters' Tea Party. In the end, you choose Breakfast Tea, a conservative hunch, hoping for a return to the unitary normality of childhood, as you rehearse your vowels over the raised saucers. Later, tea knowledge, like knowing the names of grape varieties and vintages, becomes part of a new dictionary of distinction. The medicinal properties of different teas (constipation/rheumatism/insomnia) are carefully spelled out in the half-light of a college room by Bronagh with the blue mascara. In this leafy apothecary, she carefully spoons out the dosage, the tea-leaves an illicit drug from some unnameable source. As you drink the flavoured waters and wait for the miracle of a bowel movement or a rogue snore, you begin to wonder about the benefits of higher education and dimly sense that the whole scene may be a candid-camera set-up planned by your sceptical tea-drinking ancestors who can barely control their mirth as they watch you po-faced sipping Strawberry, Blackcurrant, Ginseng. Later, there is the convalescent promise of herbal teas and their vague association with ill health and a kind of displaced preciousness. Camomile. Thyme. Verbena. The pastoral poetry of the labels gives way to careful sipping and cupped hands and the fugitive hope that sleep might be unbroken and the bathroom undisturbed.

Public houses always looked on tea with suspicion. You asked for tea with trepidation, as if it were the codename of a local ganglord or an occasional dealer. *You wouldn't be able to do us a pot of tea by any chance?* The favour was smuggled in as a half-whispered coda to a firm order for pints of the usual. The kettle it was understood was for the preparation of hot ports and whiskies and somehow to use it for the making of tea was an unspoken betrayal of its place in the publican's order of things. At six o'clock when the nation sat down to tea, tea was no longer available in pubs. There was a suspect femininity in the serving of tea in the bar, as if an unstated division of labour between home and away was being violated. But then the world and his mother started paying for water in a bottle and the taking of tea no longer seemed a delinquent activity best confined to morning and early afternoon. It now vied with espresso and cappuccino for the discriminating throats of the new customers, indifferent to the protocols of the dark ages of drinking.

From the anxious cups in the canteen of the maternity hospital to the relieved acceptance (*I'd love a drop, Kitty, one sugar*) in the house of the bereaved, the brew takes us from cradles to graves, the endless pouring a perpetual baptism of conversation and anecdote for the living. Is it something about being an island people, surrounded by water, a primal thirst, a nostalgia for some earlier amphibian state, a vague memory stirring as we empty the cans, drain the glasses, drink up the cups, the ducking-stool of liquid, our eternal punishment, our eternal reward?

TRANSPORT

FROM 49A TO 15B. FROM 15B TO 16. FROM 16 TO 7 and 8. From 7 and 8 to 3. From 3 to 7 and 8 again. And then on to 11 and 13 and 19A. This is the life the numbers came up with – the map of a life lived above ground in the city. The timetable, a curriculum vitae. The black scroll of names – Tallaght, Firhouse, Templeogue, Rathfarnham, Blackrock, Sandymount, Phibsboro, Killiney – turning through the years.

The 49 on a journey out to Firhouse and the school, hesitating between the dark cattle in the field and the lava flows of development.

The 16 taking you to university and then the changing of buses as you leaf through the unreliable yellow books of code to plot love-trails. All the routes like a modern Map of the Tender, laid over the city's distended mass. Buses you have never taken before, a dubious Casanova on an uncertain steed squinting into darkness and back again at the envelope with Pine Drive, Elm Grove, Beech Close, Oak Crescent, Birch Avenue scribbled in anxious anticipation. The window is coated in the hoar-frost of urban grime. You try to make out the names of pubs in the rivulets of clear glass. Has the conductor forgotten? The journey seems very long. Where are the phone boxes? How will you get back to your mother's house in Rathfarnham? When it is all over (just as you were beginning to get to know the stops and relax into reading), the numbers are love's mnemonic in later years. A glimpsed 33, watching the 45 pull out and the 79 swing into a corner and a name comes out from under the covers of time and the dream life of memory empties out fragments of floral wallpaper or a Simon and Garfunkel record sleeve or the sudden opulence of a full six-pack resting unsteadily on a walnut-effect Formica table.

The 3, 7 and 8 ferrying you into adult life and into town for meetings and pintings and the Every Night Fever of your twenties. The buses carry you to meetings and chaste conversations and then the afterwash of privilege sees you carried home in the pine-scented coffins of seven-seater taxis.

And, doubtless, when you are too old to drive and those around you too old to help, buses too will pick you up and drop you in the close clutch of streets that is your edition of the city. The shaking hand will wave the pass at the indifferent driver staring beyond you through sunglasses to a

night away from traffic and complaints in the lap
of his girlfriend.

We have Wedding Buses but no Funeral Buses.
The Nitelink is for the living not for the dead.
Death entails privilege for the survivors, the dark
hum of luxury slowing the traffic under the
indifferent gaze of passengers. Buses are the
province of the living, the transport of metaphor
made palpable.

There is a certain dress etiquette on buses.
Unless you are Going Out for the Night (and your
heart is pressed to the cold, damp window,
fumbling with shyness), dress is strictly casual.
Hence, the class shock of clambering onto a bus
away from the respectable snug of the private car,
overdressed for the expedition, asking the driver
how much it is *now* to go into town and trying to
remember which number goes where and what
have they done with the terminus and you
remember the moist feel of plastic against the back
of your knees as your shorts rose up under the fitful
shudders of the engine or the cool feel of the metal
bars as you rested your hands on them and how life
seemed to be an eternity of bandaged hurley sticks
and biology tests.

Buses carry their own internal maps and social
codes. The territories are chalked out with the
usual sticks of gender and age. The males upstairs.
Females downstairs. Age complicates this border.
Young women will sit upstairs and old men will sit
downstairs. The young male downstairs, a bashful

interloper, finds himself a hostage to remnants of chivalry. In the crowded bus, there is the inevitable surrender of the seat to the elderly person or the pregnant woman and the terrible moment of threatened public humiliation. The wounded pride of irate age rejecting the offer with a contemptuous sniff. Or the mine of sexism going off in loud dismissal of any suggestion of weakness.

No Standing in the Upper Saloon. This is cowboy territory. Shot glasses skiing across counter tops and the swagger of gunslingers as they stoop under the low roof to disappear from view. The unintended comedy of the syllabic stretch. The Chaplinesque "o" strutting bandy-legged into the word. Salons become saloons, pantalons become pantaloons.

The Upper Saloon was where the smokers squinted. When the mist of tobacco lifts from memory, the images are permanently altered. Memory is famously a matter of smells and, now that cigarettes have disappeared from public transport, it becomes harder to remember what it was like to sit on the masts of the 15B, wending its way into the city centre. To be a frowning adolescent once again at your appointed stool in the saloon would not mean being struck by the patina of poverty on everything but to be stung by the smoke and pummelled by the odour of Sweet Afton and Players and Senior Service.

It is always vaguely naff to sit in the front seats of the Upper Saloon. They are usually reserved for tourists or giggling children turning back in uproar to their parents smiling apprehensively from the middle rows. On the buses to the leafy suburbs, these seats provide the fairground thrill of grazing hardened wooden callouses at speed. The excitement comes from wanting/not wanting the glass to shatter and the trip to end in tears and headlines.

Single deckers are always oddly disappointing. They lack the statesmanlike hauteur of the double deckers – de Gaulles on wheels, belching darkness into damp mornings.

In the bus which went to Firhouse National School from the bottom of Ballyroan Road, the driver's compartment was shut off. He was a remote God. A Phaeton flirting with the sun or Cúchulainn in silent counsel on his chariot. The startling intimacy when the driver is no longer apart but is there in the bus behind his garden gate. Then the glass goes up again – the transparent walls of fear – and once more they are among us but apart.

The ritual of thanking the driver as you get off at your destination as if you were alighting at a windswept terminus on a remote peninsula, a distant relative in an old Morris Minor waiting for you, beckoning to you through darkening rain, outlined in the Cyclopean grope of a single headlight.

The *faux ami* of the French class. The Irish conductor did not drive; the French *conducteur* did. The French bus driver of course in the post-Revolutionary spread of privilege was a *chauffeur* or, when engaging in (expressly forbidden) conversation with a member of the general public, was a *machiniste*.

Bus conductors humming, singing, whistling. Fingers raking through the change. *Fares please*. The curious ones who asked you what you were reading. The married rogues chatting up Young Ones. The bus conductors were like nomadic agents, binding a community together, however temporarily. Then the silence that descended on the buses when they were phased out. Great Gaps in Irish Song. So Irish modernity took a new twirl on the floor of progress as piped noise invaded our bars and silence filled our saloons.

It is not just the conductor's leave-taking which has silenced many buses: sound now dwells in private caves. A number of years ago, there was a short-lived experiment, Music on Top – Ceol thar Barr. The Upper Saloon was provided with a soundtrack, afternoon radio music, middle of the road with the odd foray into folk, which had a loop-like quality, the advertisements repeating themselves every so often so, like a sinner in a medieval hell, punishment was circular, your ears roasted with the attractions of the Peak Inn in Bray every four minutes. The experiment was discontinued, presumably because even hell has its limits. Now, public music on buses, like violence, is a random affair. From the scratchy dissidence of the tranny to the tribal thump of the ghetto blaster, Music on Top means We are on Top. For teen gangs the two fingers (thumb and forefinger) twiddling with the volume button are the ultimate salute, the noisy provocation addressed to the frozen necks calculating the distance to the stop where they get off with the occasional backward scowl for the most courageous. Then there is talk radio making its way up the stairwell in surreal scraps, drivers dreaming their way through morning with Marian and Pat and Carrie and Joe by their side, concerned substitutes for the garrulous conductors now pensioned off and walking the dog on Dollymount Strand, their ears still foxed by the diesel roar of Clontarf mornings. But music is now mainly the chrome conch of the Walkman, worn by those who do not walk but sit still and watch the film of the everyday unfold from the terraces of the Lower Saloon or the stand seats of the Upper. All the President's Men and Women – secret agents of the aural with their earpieces and spider-thin black wires disappearing into bags and pockets and open zips. More bouncer than cyborg.

In the old buses, there were bells only the conductor was allowed to press. A red disc in a silver circle. The silver box with the round red bell. *Press once*. The silent invitation to mischievous schoolchildren, all knees and sudden shoves as they belt upstairs to the sound of the delinquent bells tolling for no one. Then there were washing lines running along the ceiling of the bus – floating bell-ropes, the rough jerk leading to the distant peal. The forlorn tinniness of the bell like the faint cry of the drowned. There we were, then, the passengers, occasional campanologists on our urban tours.

The bus, a ship in the ragged storm of the afternoon. Shuddering and shaking. The way passengers grab hold of the handrails and gingerly descend the steps of the stairs like tourists on a cruise ship unnerved by the violence of rough weather and a difficult passage.

Your mother's shopping expeditions to Town. The City was always a Town to Dubliners on the Edge. The grand journey. The old buses, veteran elephants from Hannibal's campaigns, trumpeting their way slowly through the thin traffic of the uncertain centre.

The complaints of the mothers. An ancestral fury whipping around the seats. Sodden maenads roaring the curse of unborn generations at It's Not My Fault the Bus is Late. The stony impassivity of the passengers and the silent mortification of the son summoned loudly to witness the national disgrace. The angry change of gears and the muttered epilogue with a chorus of scarves nodding in tight-lipped agreement. As if the buses were a grim metaphor of the failure of life in the New State, like bad chocolate and ersatz coffee in the People's Republic.

"We put up with too much in this country."

"Never complain."

"They get away with holy murder. They do."

"You're tellin' me an' I after waitin' here this half-hour."

The buses so long gone turn up in groups of threes and fours, nervous girls at a deb's dance, spilling noisily into the Ladies, staying close for the comfort of derision.

The bus tickets. Blue squares. The smeared Irish. The stricken reader getting his print fix – reading and rereading the same cramped Gaelic legalese. You take the ticket. Fold it in two with your fingers. Flatten the loose ends and press them to your lips. At the other end, a small, jagged hole turns the ticket into a party whistle. There we are, a gaggle of ten-year-olds, blowing away on the country roads to Firhouse, a great stomping riff as time tilts sideways. The taste of moist newsprint on your lips and you are reminded of the sessions years later when your face is islanded with bloody scraps of newspaper, the sad geography of a shaving expedition.

The city checks in its new inhabitants in the floating world of the bus. The arrival of the Spanish students in summer. Wolves howling for noise on the upper deck. The natives whispering disapproval. Irritation tempered by the erotic charge of these incomprehensible syllables. The girls, wrapped in the loveliness of their mahogany hair and sunsmiles – the boys, mysteriously transgressive in the fragrant veils of body oils and aftershave. The au pairs sit complaining loudly about their families and what they get for breakfast.

The immigrants in pairs seeking the companionship of language as they pick their way through the catacombs of bureaucracy. The buses are our improbable empires, frail Hapsburg vessels carrying their many-tongued population from one corner of the city to the other. And like in any empire, there are the Angry Ones, haircuts wielding the cudgels of abuse, reducing the world music of speech to the taut silence of fear.

Your friend in secondary school and yourself had created a new section of the Imaginary Library – Bus Books. A bus book was either a very large book or a small one with a very large and complex title designed to impress an audience of inquisitive passengers. Our favourite bus book was in fact more like a pamphlet, an offprint of an article by Albert Einstein you had found in the shed of an aging parish priest. Bertrand Russell and Martin Heidegger were other candidates for this theatre of precociousness. The bus book, in fact, sent up our own pretensions – the morbid timidity of puberty slugging it out with the mind's own vanity. Books are easier to read on buses than newspapers. In the packed rush-hour bus, there is the logistical nightmare of folding the page of a newspaper without dislodging your neighbour's incisors with a maverick elbow. Like eating an airplane meal in economy class with the toytown cutlery and the shallow bowls. The Tea Party in Economy. Sindy

Flies Coach. Your elbows firmly by your side, a self-important hen in the coop, brooding over in-flight magazines and the Brownian movement of chickpeas.

The Last Bus. The tumult at 11.25 p.m. and then the strange silence when they had all gone. Cursing the last gulped pint as you stared at the empty bus stop knowing the suburb was a cold trek away. Sometimes one passenger leaning into darkness gave you hope but then you realised they were too drunk or too lost to know any better. When sense took you by the hand and you made it to the bus on a Friday night, the upper saloon was Bahia, a carnival of conversation. Everything was just slightly out of focus. And then the moments of haphazard concentration. The nape of the man in front suddenly looms, steppes of bristle over corrugated flesh or his jacket picked out by the dalmatian specks of dandruff. Or else your nose tracks the scents of apricot and apple shampoos, the enigma code of perfumes, wishing you were losing yourself in that tumble of expectation and not facing into the virgin embrace of the duffle coat, tramping across Marian Park, lips tightened against solitude.

They laughed at you in the Quebec winter when you put out your hand at the bus stop. You die if the bus does not stop, they said. A little girl a driver did not see at a stop is later found frozen, wrestled from life by the Snow in her great white cloak and the North Wind, all wrapped up in furs. The Selfish Giant stares blankly into the night.

The Rush-Hour Bus. Pilgrims on our way to distant Canterburys. A Ship of Fools foundering in traffic. The worried tourists badger their spouses with maps the wrong way up and the young settle into the gentle, autistic, rocking motion of the personal stereo system. A mother is wondering how to manoeuvre the buggy through the market-day crush of the Lower Saloon.

49A to 15B; 15B to 16; 16 to 7; 7 to 3; 3 to 7; 11 to 19A.

Time's hopscotch. The first square holds the five-year-old readying himself to take flight from the moving platform, more dodo than gull. Margaret Cronin on her Black Wexford by the Rowan Tree waits for her charge and off you go, your outstretched legs a safe distance from the wicked spin of the spokes. The second square is covered in water. It comes from the sodden ends of your dripping flares. One shoe has already made a ragged hole at the bottom of the left leg. Sitting there, hunched over the future, on the thicker foam of the seventies, your mind aches for a globe of possibility, a way beyond the perimeter fences of weekly report books and class tests. The third square is peppered with the orange caps of stout bottles, the present a protracted holiday of conversation, occupations and partying. As you take the bus home from the university, you measure the distance between the communion photo on the mantlepiece and the dishevelled insurgent marking time at the Sunday table. The fourth square smells of after-shave, love anxiously drumming its fingers on the cold steel bar of the seat in front. As the glass and rubber shutters of the bus door crumple with a hiss, you step into the gardens of the night and the mingled fragrances of the future. The fifth square is the head lost in lecture preparation, calculating the distance in a mental Braille of stops, an air traffic controller of commitments, organising the daytime traffic of work, family and play. The sixth square is a slide show, the scarfed neck craning to get a better view of the city that no longer is, unmoored by the present, and memory keeps the carousel turning. The stairs now are Himalayan in their difficulty and you cannot remember where you put your bus pass. And the last square. The last square is empty. The stops have been abandoned. The nightlight shines in diesel puddles and rain hushes the heart. There are no more buses.

ENGAGED

YOUR FIRST TELEPHONE NUMBER, LIKE YOUR FIRST day in school or your first taste of blood, is the one you always remember. Nine double O four two O. 900420. As if the whole of childhood could be condensed into the six numbers written in pencil on a white card under a plastic cover. As you move out and on, the slot-machine dance of digits begins and the numbers reel until finally they stop and your own children fashion a remembered life out of the unchanging numerals of the family phone. And then they too will relinquish that shared code and the years will become an elaborate process of forgetting (the old number) and remembering (the new). 602576. Six O two five seven six. The intimacy of number. The piano fingers begin to dial the number until half-way through (a short gasp) and you replace the receiver irritated at the tenacity of habit. You were ringing the old number again. Or you start giving out a new number and only get part of the way – three four four, four, eh, four, eh – when darkness falls and you sheepishly say you'll ring with the new number.

The black, Bakelite respectability of this suburban Hermes, more substance than light. When you lifted the receiver, one end had an arched hemisphere half-covering the mouthpiece so that when you spoke for a long time the inside was moist with the confessional hiss of your conversation.

The finger in the pitch-and-putt holes of the dial and the cool, metal comma at the end. The long and short spins of the wheel. Like a familiar train rhythm, that winding and unwinding. When you went to your uncle's house in Limerick, the fascination at the telephone with the blind dial and the handle at the side you turned like a car crank. As if this was the future, not the telephone in Dublin.

The indoors version of knock and run. You take up the receiver, ring 172 and then replace it. A giddy flight up the stairs and then the repressed hilarity of the wait as the phone rings and the sharp nudge as the kitchen door opens and the receiver is lifted and the number is carefully enunciated and then there is the pause and the puzzled "Hello? Hello? Hello?" until the dark club is put back on its cradle, tinging briefly. A second attempt too close to the first was always perilous and aroused suspicion. So you waited until another reckless Saturday or indifferent Sunday narrowed the options to the call with no caller.

The couple down the road with the gleaming Ford Anglia had roses in their garden and a white telephone. The white telephone lay on the wrought-iron telephone stand with the half-moon glass top. The dial inset and the finger holes were crimson. The glacial stillness of the couple of a piece with the white elephant in the unheated hallway. A shrine to be untouched by the supplicant sores of the faithful. Only to be used in cases of intercontinental strife or the annunciation of grief (the wrong time of day, the careful replacement of the receiver, the door handle slowly turning, "Nothing serious, Tom?", the dryness in the mouth).

The yellow telephone boxes with the wooden doors and obscenities and telephone numbers carved into their frame. The half-cup handle diagonally slanted welcoming the frozen fingers, tips wrinkled by rain. The small silver cup for returned coins like a holy water fount, shiny with irritation. The dog-eared telephone books. There were two then. One for City. One for Country. The sudden stillness and the fumbled kisses in the fugitive warmth of the box, unlikely sanctuaries in the rawness of morning as the party finally spilled out its contents and you waited shivering for the first bus into town.

Press button A to speak. Press button B to get your money back. The pressure that has to be applied to the small, silver cylinders. The panic as the button resists and the distant voice shouts, "Press button A! Press button A!" And then the puzzled expression as the call fails to connect and B is depressed and no coins clink in answer. Consternation turning to frustration as the obdurate black rectangle refuses to surrender the imprisoned coins and button B is thumped with the angry impatience of a debt collector on his rounds.

The drawing-room comedies of crossed lines. Dying to jump in like a prompter in the wings with an unscripted witticism. Lovers' conversations were particularly sought after, and sisters would be summoned to eavesdrop in silent hilarity on a halting duet from Rathmines or Crumlin. The curious stops and starts of human conversation like evening traffic in the city. Afterwards the dialogues would be re-enacted on the linoleum stage of the kitchen, the tender platitudes of affection transcribed into the raucous satire of teenage cynicism (*Mary, I don't know what it is about you … Tom, was that Brut you were wearing the other night?*). As the service improved, the occasions for comedy declined and lifting the receiver was no longer a happy roulette of indiscretion.

The scratchy pop anthems of mobile phones rousing their embarrassed users to life. The instruments stowed away under an Egyptian pyramid of zips or buried under the rubble of lecture notes and overdue library books. The bus audience listening in on the withering banalities of rendezvous-speak:

"I'm on the bus."

"On the bus. The number seven."

"Wh –?"

"Wha –?"

"What?"

"Where?"

"Zanzibar?"

"Need to change. Say eight outside Front Gate."

"Eight outside Front Gate."

"Fine."

"See you then, bye!"

"Bye!"

"Bye!"

A friend who returned from Italy describes mobile phones as the final triumph of Oedipus. In Rome, for days, she listens to the young men engaged in endless, furious conversations with their mothers. The distinction between private and public breaks down. Mobile phones allow public space to become a private affair, the curiosity of others a matter of indifference, and private space is made public. There is no longer the reticence of the booth as one finger in the ear like English folk singers we bray our arrangements to the world.

Standing there in the bus silently fingering the keypad as if wielding a remote control without a set or watching the silent screen for the scrambled language of friendship or more. Not to be called is to be left in the corner of the schoolyard, outside, alone. The call for help. So the endless, needless calls telling friends where you are, what you are doing, where are you going, what you think of Kevin and Eileen and Joanne … The keys make you fidgety. The communication fix. There is the urban glamour of restlessness, the constantly ringing phone a mark of social preferment, wrestling with phones and personal organisers and airline tickets in a dream of connectedness. *Can you give me an update? The diary does not look good. Let's make it Friday.* You can tell the age of the users by the way they hold mobiles. Older users tend to hold them as if they are telephone receivers and can never quite get the dimensions right, puzzled by the smallness of their personal communicators. The young are more accustomed to the shrinking world of appliances and the toytown dimensions of late modernity. They are quicker on the draw. When a phone goes off in a room, ten hands reach simultaneously for the holsters and then there are four mumbled apologies as only one phone finds the target.

The mobiles have silenced the hallways of the blocks of flats with the dark, floral-patterned carpets and the mudslide of uncollected mail. No more the random strike of the phone ringing and the receiver lifted by an unknown hand, cradling unfamiliar voices:

"Who?"

"James? James Kelly?"

"What flat is he in? Number three? OK, I'll see if he is there."

The clunk of the dropped receiver, the receding steps and the bailiff-hammering on the door, surname giving way to instrument, "James Phone! James Phone! James Phone!"

And then the enigma of return. Would the familiar voice of your friend demand briskly, "Hello?" or would it be the breathless stranger, panting defeat into the mouthpiece, "Sorry, no reply up there."

The elaborate hide-and-seek of the flat phone. Crouched in the semi-darkness of the bedsit, radio down, television off, ears cocked for a break in the conversational hum of the hallway. Like an expectant paratrooper or a small-time criminal, waiting for the signal – the door bursting open, the dash down the stairs, the triumphant wrestling with the solitary phone. This is how obscure hatreds begin. The irritable boredom of endless postponed sorties. The tenant in flat number two who seems to spend all his bloody day on the phone to his girlfriend. The One from the States in the basement flat who never shuts up, spinning out the web of her days on Call Collect, so that not even coin shortage (there are steel strips over the 5p slots) can dam the incontinent flow of her transatlantic confidences.

The person at extension   is not available to take your call. Please leave a message after the beep. The mid-Atlantic wastes of voicemail. Or else the waiting-room ennui of John Denver's "Annie's Song" (*You fill up my …*) on James Galway's golden flute or the Vivaldi continuum – spring, summer, autumn and winter – as you wait for all the operators that are busy to come and rescue you from the Desert Island discs. The curious conversations by proxy as answering machines chase their virtual owners around the city of bits. And the messages on the answering services in the halting diction of patients in intensive care, phrases punctuated by a baffled absence of breath, "You [pause] have [pause] two [pause] new [pause] messages [pause]. To [pause] listen [pause] to [pause] your [pause] messages [pause] please [pause] press [pause] one [pause]."

The larger the family, the more valuable the phone. Waiting on the landing to swoop and then fending off irate parents (the cost) and annoyed siblings (the time). The indignant dumb-show of doors being opened and watches being pointed to and promises to ring friends being mouthed in angry slow motion.

The phone ringing in the empty house. Something infinitely desolate in that sound. In films, actors are always woken up from sleep by phones, the women grumpily massaging the roots of their hair as they ask Who Is It? Downstairs in the kitchen, the cordless phone seemed an extravagant icon of progress in television serials with men in expensive shirts stirring cereal bowls endlessly as they spoke deal code into a phone clamped between mouth and shoulder. As these Quasimodos rambled around their stadium-sized kitchens, it was hard to know whether they were grimacing with the effort of holding the instrument (as telephone engineers primly put it) or were simply wearing the conventional rictus of television normality.

The telephone accent. In the kitchen, language was at the North Kerry races until the phone rang. Your mother walked decorously to the phone, her face taking on the expression of constipated seriousness she wore in the more solemn moments of the mass as she lifted the receiver and repeated in grave stillness, "Nine double O [long pause] four two [short pause] O." The telephone accent was a private version of posh, cobbled together from years of emigration to Birmingham and the starched speech of Sunday matinées. When you rang home you were first presented with this baroque diction and the theatrical pauses and then as she recognised your own voice you could almost feel the neck muscles slacken and the mouth break into the easy canter of familiar words. At the phone, in that initial moment of contact, she was both maidservant and duchess, deferential and remote, the telephone receding from modernity into some distant comedy of manners. As if sitting at the telephone table she should have worn an elaborate wig or had the rouged cheeks of a society belle. Older people still answer the phone with that particular mixture of hauteur and caution. As if the news (tests, accidents, entitlements) is always bad.

The clinical cheeriness of Convenience Speak – "J.V. & C. Products, Catherine speaking" – with its codes of evasiveness. The brisk putdown of "He's in a meeting", the conciliatory, "Can I get her to call you back?" or the ominous, "Does she know what it's in connection with?" Occasionally, you are left wandering in silence and, feeling foolish like the stooge on a candid-camera gag, you mouth, "Hello? Hello?" the second hello always slower, more hesitant than the first as doubt begins to seep in and you realise that the receptionist like the textbook neglectful parent has left you in a cloud of unknowing with your own sceptical voice for company.

The Christmas-stocking addendum of the Address Book. These books like socks and strong-smelling bubble baths are affordable for young budgets. There were not many numbers to put in the early books: your parents; your best friend; the guy you met in the Gaeltacht. Usually they ended their brief careers as scribble pads for afternoons adrift in winter listlessness. The Dadaist collage of the first childish entries – the large P, small a, larger R, smaller e, tiny n, larger (still small) t, huge (relatively) S. Then the unstable boundary wall of the full colon, one stone breaking free, the other unsteadily remaining. Then the number itself, printed like a music store, all hops and dips and slumps.

When the numbers did multiply and whirl about your life uncontrollably, the notebooks were no longer to hand. The margins of bus tickets and the flyleaves of paperbacks or the borders of personal notices in newspapers were covered in numbered ink trails. The consternation over toast (stale bread) and scalding hot tea (milk carton empty, not in a fit state to go to the shops) at a number on the crumbling edge of a beer mat. JOE 2733596. Joe? Joe? You sift through the debris from the night before, go through a head-splitting roll call of friends present but Joe seems to have been erased from the historical record. As your eyes flicker over the contents and vitamin count of Corn Flakes for the thousandth time, Joe fails to register in any of the searches, his number a scrap of Linear B you have failed to decipher. It is only later as you tease the groceries out of the basket under Linda's sullen stare in the local Spar that a shard of memory shivers for a moment and yields the code. France. Joe loved France. He had been there many times, still had a French girlfriend in Poitiers. In the false intimacy of drink, a shared passion ended in the fumbled exchange of numbers and a promise to ring that seemed both genuine and urgent at the time. Of course, you will never ring Joe and Joe will never ring you. Like the Eurorail addresses jotted down in hostels in Barcelona, Prague, Florence that might make it to the afterlife of a postcard but generally lie where they are written, frozen reminders years later of a damp night in an overpriced city or of the dutiful logging of national differences in the pidgin speak of the breakfast room (*Yes, in Ireland we like music*), lust simplifying the tenses.

Careless Talk causes Hassle. In the intelligence wars played out between adults and teenagers this is the unspoken motto of the young. Mono-syllables (grunts) protect information sources and limit knowledge of movements. Sitting-room armchairs become listening posts in the low intensity operations of family surveillance, the parents like anxious cuckolds trying to unscramble the code in the hallway. Counter-intelligence is not made easier by going mobile. If information can be exchanged more easily, the weight of subterfuge is not lessened, only transferred.

"Why did you not answer the phone?"

"We were worried sick."

"Where were you until this hour?"

"What do you mean the battery was flat?"

"What's the point of our getting you a phone if you always have it switched off?"

"The what? The network was down! Jesus wept. What do you take us for?"

I hear you calling me. It is on these telephone lines and in these networks that the Bildung-stories are written. No more breathless letters to young men in breeches and young women in gowns, using the light of casement windows to make out the copperplate rhapsody of enthusiasms and affections but hours of talk dwindling into oblivion, the peculiar mixture of confession and malice that always makes for good conversation, in the snugs of our hallways and bedrooms and kitchens. There is no Socratic preparedness in the *agallamh beirte* of the phone, more a kind of endless jamming which ends in the theatrical Is It That Time Already? Phones have their Bogarts. Unhappy users who prefer the clipped prose of instructions and the minimalism of the mono-syllable to the verbal meandering of callers delighting in the riverrun of contact.

You live in the Call Capital of the World. Young people in Dublin telecommunications hatcheries listen to the complaints of continents. There they sit in Babel's echoing ruins administering to the unhappy and the incompetent in all of the languages of the Corporation's paying customers. The New Vienna is a Call Centre. Languages clamour here in the scripted therapy of the Customer Sales Representatives. And yet how vulnerable we are in the dangerous proximity of telephones. The hotel reservation made in the foreign language like ringing a former lover, the heart threshing and the same phrases idiotically repeated over and over, and then the flame burst of words from out of the midst of a Venetian afternoon or a brisk Berlin evening. Hostages with a simple message for foreign humanity and the assault rifle of necessity in the small of our back, we stammer out the simple sentences. Our accent comes back at us, magnified by the receiver, as if we were forever doomed to be Comic Chappies trying to speak the Damned Lingo. The hilarity or bewilderment gives way to the humiliating climbdown of Basic English (*Hold. One moment. Please. I get someone*) and the clammy receiver is replaced with the pledge to do more work on listening comprehension (never fulfilled). And you think of all those polyglot spies listening effortlessly to the guarded chit-chat of the Cold War or the intelligence officials in England listening to the torrent of subversive talk from Irish lovers through all the years of the Troubles, wondering did they have any language problems and what must it be like to tap the phones of the most garrulous island in Christendom?

There are other numbers you never forget. Or rather, you never forget the voice that answered the number. The numbers you do not put a line through in the address book because, unlike the others, there will be no new numbers above or below. The blue, red, black alpha waves of change, the seemingly endless, biroed annotations of the living stop. On the giant notice-board in the black vastness of the station the spasm of clicking subsides. The numbers come up. 900420. Bingo.

THE INTERNATIONAL BRIGADE

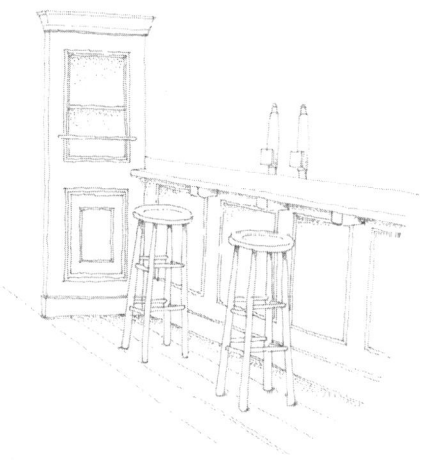

LADS AND LASSIES! TIME NOW! LADS AND LASSIES!
The call still echoes although Paddy is no longer
there, his Maypole invitation to waltz out into
night replaced by a coat of arms over the bar. This
is a bar which is almost too big for the Bar. A Trevi
fountain hemmed in by pizzerias. The door from
the serpent-twist of Wicklow Street opens onto the
great wooden altar of the International. Like a
complicated Eastern temple or the hallowed site of
an Afro-Brazilian cult the ledges and recesses,
nooks and crannies, glass and polished wood are
freighted with the unholiness of the kitsch and the
everyday. The drinkers on the high stools at the
marble counter let their eyes ramble through the
peanuts and Friendly matches and seaside
postcards and the wooden breasts that stretch into
the posed paws of the eternal Sphinxes hovering
above eye level. The impatient wait for the friend
(*He's always late*) or the anxious tarrying for a new

girlfriend (*Did she say eight or half-eight?*) fills those
watchful eyes with the forensic thoughtfulness of
the archaeologist, scraping through detail. As your
gaze wanders through the glazed detail of the
mahogany labyrinth, itemising the bric-à-brac,
time expands and slows as the stout fingers the
tendrils of memory. The narrow bar stretching
monumentally from door to door has something of
the Victorian railway carriage in its yellowing
afternoons and from there you are transported to
the other holy wells of the city, drinking in the
shattered refractions of times past.

O'Neill's. The Stag's Head. The International
Bar. Burke's. Peter's Pub. Neary's … The name
cards are shuffled differently for each drinker in the
city, a private Monopoly board of association and
recall, pubs marking out phases in existence,
girlfriends, boyfriends, politics (Provos drink here,
Sticks drink there), education (UCD medics there,
Trinity jocks here, DCU communications students
elsewhere), a new bed-sit, the first job. If directions
to foreigners are often given using pubs as
landmarks, the landmarks are not just public but
private. You could imagine a biography that would
ignore the usual chronological minimalism of
1954–1972, 1973–1993, 1994–, or the homely
platitudes of "The Early Years", "Growing
Awareness", "The Prime", "The Twilight Years",
and have a table of contents that would read
Chapter One: *Delaneys*; Chapter Two: *The Buttery*;
Chapter Three: *O'Neill's*; Chapter Four: *The Stag's*

Head; Chapter Five: *The International*; Chapter Six: *Peter's Pub*; Chapter Seven: *Neary's*.

There are lives where there is only one chapter heading. One pub. One life. The Local that is the universal, which carries you from the first public drink through the twenty-firsts, engagements, weddings and christenings to the birthdays, anniversaries and funerals. For the locals there is the imprimatur of the free pint at Christmas time but your own loyalties were too fickle or your life in the city too shifting or unstable for the ready recognition of the cocked eyebrow, the affirmative nod and the pulled pint. Instead, there was the succession of Locals, each holding a fragment of your story and there are of course the unfinished chapters, the pubs that claimed your attention but never kept it or that remain in a no-go area of the heart, cordoned off by loss.

There are bars like sitting-rooms, where you sit and stay and stare or talk. And there are other bars, which are bars of passage, where you stand and watch or talk and go. This bar is undecided. The two great doors at either end are invitations to the busy exit and entrance and standing noisily *en route* to other business is an accepted custom. The high ceiling and the turning fan and the ceaseless passage is the exotic dimension to the pub, its other life as a café-society sophisticate. But for those who remain, the doors are sealed, the narrow bar a large room where conversation deals its hand endlessly.

Pubs have their hostages. The big drinker in

college who somehow mistimed his exit as he
trundles from pub to pub in search of company.
One minute, it seemed everybody was there
roaring for more and the next they were postcards.
The signatories would pile in at Christmas and mix
condescension with nostalgia as they asked him
was he still drinking in those same places. As
students, they delighted in the social promiscuity
of the pub, the unstable democracy of the barstool.
This was before qualification, career and marriage
led to the raising of the drawbridge and the retreat
into the tower houses of privilege where they now
entertain, fortified by vintage bottles and choice
cheeses. For the Departed, the memory has been
archived, the photos sorted, but their past is the
Drinker's unchanging present.

There is the Pub Friend. You only ever see the
pub friend in the pub. The conversation is always
public and satirical, never personal or confessional
(except in advanced and immediately regretted
states of drunkenness). Jokes flourish here, newly
rehearsed, topical and, as you pull away from the
nightly orbit around the pub friend, you feel the
humour leaking from your own memory, the odd
joke haphazardly remembered and repeated over
and over (*Did you ever hear the one about the faith
healer in Athlone ...*) to your exasperated partner
and eye-rolling children (*He's not going to tell that
one again*). Here is the peculiar social intimacy of
the pub, the face you will see for years but never
the house or the flat where pub friend lives, as if

the real parlour was here on the motley tiles of a city pub and not the weekend-paper strewn front room of a bed-sit in Rathgar.

Downstairs in the International it is mid-week and the courting couples, hands clasped in front of half-emptied pint glasses of lager, hold up their childhoods for mutual inspection. The other tentative hand on the inside of the thigh is warm in the complicity of remembering. Now that the groups have dwindled to couples, the glasses are lowered more slowly and there are fewer jokes, as talk turns to How my Mother gets on my Wick or How my Oul' Fellow is always Wanging on about the Government or else it's spluttered laughs over the Terrible Times in Tramore. The tapes of early experience have a long time to run, so there will be many more of these hand-holding confessions before conversation becomes more code than revelation. Then as the years pass, there are the degrees of departure. People begin to move to more distant suburbs or faraway cities. They begin to complain about the smoke. And the noise. And that you cannot find a seat. The evolutionary regression of age, from *homo* and *femina erectus* to the seats of the established. Soon it will be wine rather than beer. The dinner party finds favour as jostling with youngsters begins to exhaust the spirit. You will of course bring your foreign visitors on a tour of the pubs but you are now as much a stranger as they. Wondering should all these people be served? Are they not under-age? Forgetting what

you looked like at nineteen. Or on your holiday in
the West you will sit in the late afternoon with
your wife and restless children, briefly envying the
monuments of ageing intellect at the bar and
feeling that old, familiar languor return as evening
pushes shoppers in the door. But the kids have
finished the Cokes (minerals your mother called
them) and scattered the crisps and any longing is
disciplined by the terror of taking the wheel on
dark, pockmarked roads.

On the ground level, the seats had collapsed.
Sinking into them was easier than pulling yourself
out of these velveteen dugouts at the end of an
evening or when your bladder was threatening to
burst its banks. The Depression. This is where you
sat out the documentary years of Southern
unemployment and Northern violence, the city a
host to the tumbleweed survivors pooling resources
to clasp forgetfulness in Sides or Suzy Street.
Conversations rambled through the evenings to
reach the same weary conclusion, that the country
had once again fallen and that the only virtue of
corruption was gossip. The streets outside the pub
were thinned by exile, and inside the terrible
truculent ghosts of the fifties would sneer at you
from the bar if the evening failed to pick up. And
then the World Cup spilled into this stunted
world. The Romania match and we are standing
on the seats. On the floor, a man from Westmeath
is locked in prayer, his bowed head a still point in
the sacramental silence of the penalty kicks. When

Packie Bonner breaks ecstatically from the goalmouth, it is more jailbreak than victory burst and we the inmates come pouring out after him, sensing some impossible spring of change, anything to clear the brown fug of emigration and deceit from our tired lungs. In the world beyond the stately swinging doors of the pub the Guildford Four are still in prison and, along the quays, post-match revellers and marchers in their dark overcoats merge and O'Connell Street turns into a mass of Dubliners swimming in their own disbelief, riding the bumpers of Volkswagens and Fiats like drunken charioteers returning from campaigns to be remembered in the half-light of a sore morning.

The peculiar time-motion studies of the pub. This is apparent in getting down off a barstool. Like mountain climbing, it is always much more dangerous going down than going up. The drink slows time down but expands space. It now seems a long and treacherous way down from the top of this unsteady Olympus. You touch solid ground with the tentativeness of a novice ballerina. As the evening wears on, the metaphors, however, become more Spatial than spatial. The ground is now in orbit and touchdown demands timing, precision and luck. When you do land, there is the slightly self-conscious moonwalk to the Gents, the unsteady tangents and the sashaying hips as your head peers through the helmet of One Too Many at ordinary life back on earth. The Gents itself is a reduced affair. Two peeing stalls and a toilet with a door that never locks so that there is always a poor unfortunate on the bowl bent forward with his forlorn foot and rigid right arm held up against the door. At face level in the stalls, Gentlemen are met by an iron grille and the omnipresent smell of bacon fat. Conversation is rare early in the evening as the sober head stares coolly into the meshed darkness but later circumspection fades and there is the jokey banter of Gents on the Town – swaying conversations these, for all the world the men like drunken parrots on a perch. Sometimes in the absence of talk there is the intense species of concentration that comes with the lowering of drink and the passage of time and two words on the porcelain white of the urinal begin to fuse into a giddy mantra of possibility – TWYFORDS ADAMANT.

The words begin to fall into sentences … The Twyfords were Adamant that Priscilla would never be admitted to the house. No, not under any circumstances. Jane was quite firm on this and George agreed. Priscilla, after all, was rumoured to be related to the Armitage-Shanks, a family whose profligacy was matched only by their sneering indifference to the conventions of polite society … The Twyfords-Adamant were one of the oldest and most respected families in Boston. Where their money came from, nobody quite knew and, because nobody could remember, it no longer mattered. What mattered was that they had it. Lots of it … The Twyfords and the Adamants saw much of each other that summer. It was a late Edwardian summer, all parasols and plimsolls, and even Aunt Lydia agreed that she had not seen the like since the year of the Prince Consort's death. Bicycle bells jangled and smoky cars roared in the still afternoons of early July … The stories are abbreviated by the arrival of new disciples of the Divine Words and it is time to leave the stall to other meditants.

Tourists sometimes make their uncertain way into the International Brigade. They wear the uniform of their age. Bright-coloured, multi-zipped, Gore-Tex rain gear and walking boots with new shoelaces which always look incongruous in this most urban of pubs. They nurse a single glass of stout (an experiment) for hours and stare bemusedly at the locals as if waiting for something to happen. Nothing does. Only conversation. And more of it. Most of it incomprehensible to ears diligently schooled in the Upper Form Prefect English of continental Europe. Eventually they leave at half-past ten and wonder what all the fuss is about. There has been none of the music they see advertised in the brochures and the only words they can make out are "same again". Same Again. A friend in college wanted to write a play with that title or maybe he did. It was to be a familiar tale of Irish stasis and quiet desperation in the rain. Presumably there were two characters, both angry, both unhappy, waiting for the train to pull out of their lives but never getting further than the Station Buffet of tipsy speculation. Same Again, Simon.

Some do stay, of course. The first, wild year in Ireland on the grant or the language assistantship. A great spill of pints and parties as the tongue makes its peace with Irish vowels and the jokes come into focus. The easy novelty of foreignness as desire homes in on difference. Since no one can place you, you drift from the caravans to the mansions, equally welcome, equally outside. You transpose what you see to what you know and realise that in Vienna or Paris or Berlin these doors would never pull back to let you in, your accent a giveaway, your language a marker, an object of suspicion until your coordinates have been sorted and you are fixed on the social grid (cop, social worker, parvenu, hustler, Bad Influence). A bottle of whiskey makes you sick but you start drinking those huge pints of beer and stop telling your friends that you're not thirsty when they ask you do you want another drink. You find the words of a song you learned in primary school and this becomes your party piece. Not being able to sing was not the point after midnight. And the song always worked because nobody understood the words and the pitch was unusual. Years later, settled in a suburb of the city with your bilingual children, mistaken for Irish by the English friends of your mother, you wonder how you managed all those nights in the pub. You only go there now after a Sunday stroll on the beach with your husband, the software engineer, and have a coffee (much better now than when you first came) or sometimes an orange juice. Sometimes your in-laws have gatherings in the pubs but you are deafened by the noise and the smoke makes you cough and your clothes smell for days afterwards. Your husband is the New Irishman who does not like pubs. He reads wine columns and takes elaborate care in pronouncing the names of the wines he buys. He never quite gets it right but you have not the heart to correct him. He thinks pubs are for Young Ones and Drunks, beginnings and endings. Not for him. He cooks complicated meals every fortnight and you have your friends over, mixed couples like yourselves, who complain about service and prices and traffic in the city and admire the bread. You are relieved he is not in the pub every night but his friends do seem to get through a lot of wine.

It is late afternoon in the pub and the city is awash with summer. Here is a seat for a Cardinal, quill raised, looking towards the window where the soul of his friend has shattered into daylight. There is the stillness of forgotten Sundays and the pint glass is no longer cold to the touch. And here you are, one of noise's refugees, an asylum-seeker, looking for a sanctuary of quiet in this afternoon adrift.

L'Internationale. The British and Irish Communist Organisation used to advertise their meetings upstairs in the International. So the name merged into legends of conspiracy, hungry young men hunched over the future, pencilling in the proletarian utopia in a room queasy with the smell of bottled gas. The bored Branch Officer downstairs doing out the crossword puzzle of the Far Left, dreaming of kitchen extensions. Later, the room would host the Comedy Cellar, the upstairs basement a wry footnote to the collapse of hope. And then you learned that the name was nothing to do with a forlorn outpost of the Fourth International but was named after the Five Nations Rugby Internationals by a rugby-sick owner, a companion to the Old Stand on the other corner. Yet it is not car coats that have filed through here over the years but words and their hostages – plans, ideas, broadsides, blueprints. The broken dykes of talk, sentences spilling over the glasses and arms raised at midnight in a riotous postponement of agreement.

They take a twirl here, the ghosts of our witness. And you can barely make them out. The ruddy-faced bulldog, sniffing over his pints for years in obscure disgruntlement. The knife-thin cardsharper in the leather jacket with nervous fingers and septic eyes. The elderly actress with the awkward lipstick lines and a throng of plastic bags assembled at the base of the stool, the fingers of her right hand relentlessly caressing an eyebrow. The faces snapped in a head's turn before night filches them. Etains who step in momentarily and then we are left, perplexed; Midirs caught in the delicious slipstream of that wing beat. The present is a bully of course and we throw out the negatives to keep up with the press of the here and now but sometimes in the slowed-down parenthesis of the late-afternoon pint they introduce themselves, all those who have sat with you, held your hand, pointed at you, shouted at you, kissed you, pulled the trip-cord of laughter, they squeeze in beside you on the seats tilting backwards and your mind tumbles back through the pubs of the city, through the bright, infinite circuit board of remembrance.

SHORT BACK AND SIDES

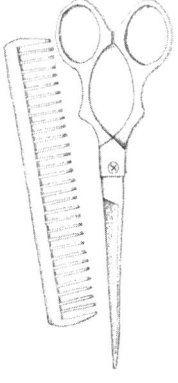

SHORT BACK AND SIDES. NOT TOO SHORT IN THE front please. These were the only phrases you knew in the language. You repeated them as you leafed through the torn, bulky tabloids. The comics, if there were any, had already been stolen. The waiting rooms of doctors and dentists always contained a slight suggestion of self-improvement, the back issues of *Time*, *Newsweek* and the *Reader's Digest* intimating that the expense of the visit would be amply rewarded by a broadening of the mind. No such pretence in the barber shop. It was the locker room without the team. Everyone from seven to seventy reduced to a common denominator of maleness, flinty stares and the restless snap of the newspaper pages. Nobody much talked to anybody else unless it was to ask for a paper or to make a half-embarrassed request for the time. As the Boy Jekylls mutated into Teenage Hydes, with hair growing in all kinds of funny places, there was much cracking of fingers and anxious, hunched, autistic rocking.

It all begins on the stool with the foam spurting from a kitchen-knife wound. The neighbour four doors up is a police detective who not only has a pistol with a wooden handle but also has a barber's electric razor. Maybe his father was a barber or maybe the razor was a quixotic souvenir from a successful piece of crime detection in his youth: you never thought to ask. Having this weapon with its thick black flex seemed to confer great distinction on the immensely tall Garda. You sat in pharaonic stillness, your anxious six-year-old hands firmly gripping your bare knees as the airborne drone buzzed noisily around your ears. There were the moments of panic as your ear was pressed back and the cold metal was pressed against the bone, mowing down the fugitive hairs that hide out under the fleshy canopy of your earlobe. Here the razor becomes a drill, the irritated whine of the tiny motor rising and falling as if trying to breach the fortress of the skull and get in and do some real damage. The nervousness never quite leaves you so that years later when you feel the cool pressure of the baby strimmer in the barber shop you wonder will it all end in tears and bone splinters. The same uneasiness that tracks the flight of the scissors in the barber's mirror, fearing a moment's inattention and a bit of your ear will get snipped off with the same brutal precision as your hair. When it was all over, and you stood up straight on wobbly legs to the automatic approval of the Garda-barber and his admiring wife, your hand went to caress the stubble of newly cropped hair. A strange pleasure feeling the spiky toughness of what was left behind. And then when you went out to play with the detective's son, the wind running its icy fingers along the nape of your neck, there was the knowledge that you had lost something, woolly jumpers and prickly collar-labels a scratchy reminder in the weeks ahead that you are now exposed and vulnerable.

Placed high up on the chair on the board covered in black plastic, you could feel the bottom of your thighs squelch and as your torso was covered in the giant, grey bib you mouthed the phrases over and over, *Short back and sides. Not too short in the front please*. The sudden elevation was unnerving and your concealed hands gripped the board as if at any moment the plank placed across the arms of the barber's chair would metamorphose into a ski slope or storybook rapids and you would drop from an inconceivable height to meet a certain death. The fear was tempered by curiosity. Set into the wall was a shelf with a glass flap. From its brilliant white interior the barber would extract a scissors or, occasionally, a comb. This you later realised was some kind of sterilising unit but at the time it seemed to be part of the paraphernalia of space travel, as if you expected to look up and find the barber dressed not in a dark housecoat but in tight-fitting intergalactic hose and starfleet top.

The trek to the stars seemed to continue when your mother went to Joan's Hairdressing Salon. When you went in there, under exceptional circumstances, it was the violation of some other, secret place. Inside, your mother and her friends were aligned in a row under a set of head dryers, leafing through magazines and shouting at each other and at the hairdressers over the noise of the dryers (*Carol, I saw you with Jim at the Do in the Green Isle the other night*). You were introduced to Joan and your mother's friends and they nodded approvingly and the words "big" and "grown" would make it past the wall of sound. The scene was vaguely alarming as the dryers looked like the cream-white helmets used for brain transplants in the low-budget science-fiction films you watched on the television. There were no coils visible or German accents directing the operations but some part of you wondered was your mother's brain being switched for that of an extra-terrestrial changeling or a lower-order primate. The way Joan and her assistants fussed around these austere cones did little to reassure you. Still, nothing much appeared to have changed when your mother returned from the hairdresser. She did not move her neck in odd ways or ask you to dry the dishes in the metallic, voicemail drawl of the cyborg. Only her mood was different. A brightness was abroad, the pampering of the salon softening her words and making her buoyant on the promise of glamour and the forthcoming Do. And then the lost world of curlers. Neighbours popping in, their hair decked out in tubes. In the drawers in your parents' bedroom, the curlers lay in readiness for the summons. Taking them out one by one, you sized up their potential not as preludes to dinner dances but as props for winter games. The curlers with the big pink sponge held in place by a white plastic clip so that they looked like some psychedelic steam roller in the Land of the Wee Folk or the blue curlers that resembled tiny plastic cheese graters ready to shred toytown thimblefuls of cheddar. The brown hairclips that were made to do the splits so that they could be used to pick locks or clean nails. The nets cast over the hair to trap these tubular fish as they settled into the long wait before evening and the release into the coiffed perfection of the night.

When times became hard, as they did with increasing regularity, it was decided to purchase a stand-alone hair dryer like the ones seen in professional hair salons. In our version of home economics, it was reasoned that the initial invest-ment would be recouped through the savings on visits to barbers and hairdressers. Removed from its box, the dryer was indeed a passable imitation of the real thing though it looked curiously out of place in the corner of your parents' bedroom rather than standing guard over nets and curlers on a hair-strewn floor. Oddly, in view of its featuring largely in hairdressing salons rather than barber shops, you were selected as the first guinea pig.

There was a thin, vertical button on the front that you slid forward for maximum heat. The difficulty was adjusting the height of the dryer, a small piece of metal needing to be depressed very firmly to raise the elephantine head of the dryer helmet from the base. Your hair was duly washed and you were solemnly enthroned on an orange stool with strict instructions to sit still. Through the movable plastic shield that was an extension of the helmet, you could make out the forms of your sisters, a chorus of barely restrained hilarity, waiting for operations to begin. The helmet was adjusted, the plug was pushed into the socket and the temperature was raised to maximum to ensure rapid results. Your nervous if expectant smile flickered and soon faded as you felt the crown of your head grow hotter and hotter. The unmistakeable smell of scorched hair indicated that you were less the Dauphin, the Infant God to whom great things would come to pass, and more a dismal heretic, destined to be incinerated for violating the Holy Writ of the Hairdressers' Guild. From inside the furnace of the helmet, like some astronaut sailing too close to an alien sun, you desperately mouthed instructions to ground control, asking to be instantly relieved of your command. The chorus at this stage was prostrate with laughter while ground control with much *friggin'* and *blitherin'* tried to raise the halo from hell from your singed pate. Eventually, you ducked out howling from the infernal apparatus, a mutant child of Joan, the botched outcast of an experiment gone terribly wrong.

There was the time when barber shops went the way of mass attendance and polite visits to the relatives: you simply stopped going. As you grew into long pants, school involved sullen duels with year deans to see How Far Could It Grow. The endlessly repeated observation about how *You can never tell a fella from a girl these days*. The great bramble growth of hair and occasionally the comb, a reluctant rake pulled through the long, high grass of revolt. Parting the hanging curtains of hair to light up a cigarette or sup a pint. Taking out the tangles was a solitary pleasure indulged in over the Sunday newspaper and when love came to town the untangling was a giggly prelude to the locking of mouths and the caressing of soles. In the great tumble of youth, you never imagined that one day the hairs might congregate ominously around the plug-hole in the bath or that the white thread brushed off the shoulder of a jacket was not a thread but one of your hairs. The speckled constellation on the shaving foam would be further evidence of the greybeard on every tide. So you are back in the barber shop, lower this time, a baffled Narcissus watching the wet sleet of cut hair fall onto the white smock that now begins to remind you of operating theatres and the routine banter on the trolley before the doors swing open, the scissors snipping and snipping through the years.

There was a period when you decided to go to

unisex hair salons. It was of a piece with your opposition to single-sex schooling and the dance-hall boundaries of Irish life, boys on the one side, girls on the other. Crossing the lines, however, was not as easy as your politics made out. Not having an appointment was the first mistake. Not remembering who last cut your hair was the second. You were then despatched to a waiting area with soft chairs and magazines and as you flicked your way through the sixth glossy, you slowly realised that only you had not moved in the last half-hour. On the salon floor a complicated choreography of busyness gave you no clue as to when you were likely to be released from the holding cell. When you were finally shown your way to a seat, there was the problem of conver-sation. All around you, long, animated discussions were shadowing the delicate manoeuvres of rinses and colours and perms but the mirror only offered you the spectacle of your own speechlessness. You generally failed on the first question, *How do you want your hair?* There seemed to be no language for what you wanted apart from a general wish to have less of it. So as if the language you had spoken since birth had become a poorly mastered foreign tongue, the result of some horrendous accident, you made vague halo gestures around the top of your head and mumbled something about *Taking some of the weight off.* A slight crease of the hairdresser's forehead and an imperceptible shrug would send the scissors into action. The spirited

voices to your right and left seemed to highlight
the high solitude of your Carmelite enclave. Your
mind would search for some topic that might ease
the log-jam of soundlessness. As the hairdressers
always seemed to be impossibly young (and
startlingly small), there was always the fear that
whatever you said might come across as the clumsy
chat-up line of a mildly psychotic swinger. So you
went for *You must be busy today* or *There seems to be
a lot of people about in town* and they would almost
invariably nod, like bored taxi drivers who had
answered the same question a hundred times about
how busy town was that night, and say *It's not too
bad* before pushing your head firmly downwards
with splayed fingers.

One day you get trapped in the looking glass of
language. In the small Burgundy village the barber
chats away and lets his curiosity draw you into
conversation. Buoyed up by your recent fluency,
you get carried along on the vanity of your own
performance, a potted history of Ireland keeping
time to the energetic cutting of the scissors. He
cuts and cuts. A smile. Nods. The words keep
coming from a mouth set free from the beginner's
embarrassment. He cuts and cuts. The mirror
offers only the image of damp sea wrack, your hair
plastered to your skull by the preliminary wash.
Another question and the sentences are out of the
traps again, detailing the scar of Ulster to the ever-
smiling barber. Cut. Cut. Occasionally, when you
glance at the feathery heap of curls rising from

your lap, there is a very faint tremor of unease but
trying to explain why Ireland is the way it is soon
has all your attention again. It is only when the
hairdryer is produced and exhaustion gets the
better of your monologue that the scale of the
disaster becomes apparent. You were Victor
Frankenstein gazing in horror at the unblinking
yellow eye of the Creature or Doctor Jekyll
shuddering into difference, appalled fingertips
stroking the hand that had become a paw. In the
looking glass was the new boy in the Accounts
Department, a solemn crew-cut soldier of con-
vention, the exposed ears an early warning to
pinkos not to come troubling him with their
shoddy clothes and phoney ideas. The unsmiling
young prig sitting opposite was you. The barber
who had struggled with the overgrowth of matted
locks circled your head with a round, hand-held
mirror inviting you to share in his triumph.
Already, you were beginning to wonder how you
would live with this Double. Would you spend
days on end in darkened rooms or slink around
under an improbable woollen hat (it was mid-
summer) or, alternatively, embrace the new you
and become a daily communicant and start
working on a golf handicap? Crashing out into the
morning light from the laboratory of your
undoing, you desperately ran your fingers through
what was left of your hair, hope cheating with
experience, trying to convince yourself that the
hair would grow back again in no time and that

you would be soon rid of the terrible twin gazing out at you from shop windows and the wing mirrors of parked cars. The bad hair day was only beginning.

The wash-basins with that neck-rest. You place your head tentatively on it, always too far up so that you have to gradually lower yourself into the helpless slouch of the shampoo victim. With your neck wedged in the groove your thoughts turn unhappily to the guillotine or the stocks of a medieval town on market day, expecting the hiss of the blade or the dull smack of a rotten turnip. A world turned upside down, with the disgraced aristocrat or grovelling brigand meeting his fate on his back and not on his knees. The voices that seem to come from the far side of a waterfall, asking if the faint drizzle is *too hot* or *all right*. The strangeness now of fingers not your own massaging your scalp and then the sudden, cold splash of the shampoo. And you are back, kneeling this time, on the chair in the kitchen in Templeogue, your eyes shut tight in terror at the potential seepage of shampoo. Random drenchings are followed by vigorous rubbings and in this sightless monsoon you wonder when will the towel arrive so that you can come up from this watery kingdom of swirling suds and lemon scents. Sitting back, a midget boxer with the towel draped like a cape over your relieved shoulders, you brace yourself for the hard plough of the comb, dragging its teeth through the crossed lines of your hair, the sharp ends marking out drills on the tender scalp. And then you bolt, dryness the wind's doing, wondering how long you can put off another descent into the kitchen sink.

The upturned cigarette, a giveaway. An affectation of the period in the dark studio photograph, a distant suggestion of bohemia and abstractions. The hair is a tight gathering of black curls, an oblique tribute to the Mediterranean in the minglings of family history. There would be no later photographs to record the contraction, the bared temples, the random snowflakes of the years. He never got to feel the chill of autumn on his scalp or to run his fingers through the thinning canopy of middle age. In her last glimpse of him before they put on the lid, your mother remembered that he still had a *fine head of hair*. For months after his death, his comb lay on the window-sill, tiny arcs of black hair caught in the teeth, lingering mementoes of growth. And when you stare into the barber's mirror it is his future that you catch in the reflection, a vision of what might have happened if the years had not trailed away from him so early. And now it is not starfleet but one of the Fates standing behind you, silent, her hand raised for one last snip of the scissors.

AUTOGRAPHY

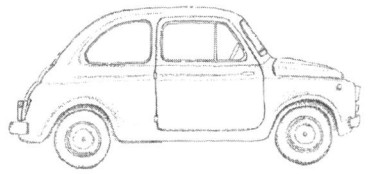

*ERRRRRRRRRRRRRRR ... ROOOMROOOMM ...
errrrrrrrrrrrrr ... eeeeeeeeeeeeee!* Stretched out on
the floor, your child's head lying in the dust bowl
of the sitting-room carpet, you continue your
noisy, urban monologue through the dark winter
afternoon. Your right arm is manoeuvring a
Batmobile, James Bond's Aston Martin and a Mr
Whippy van into a traffic jam. Down there, with
the dust of the carpet in your nostrils, everything
was possible – parades, riots, catastrophes. The
soundtrack brings drama to the long, winding line
of Matchbox autos and Dinky cars, mostly the
chance pickings of Sales of Work and School Fairs
so that the Batmobile's blade refuses to flick open,
James Bond has long since been ejected from his
ejector seat and reported missing and Mr Whippy
remains a perpetual prisoner of his van where the
doors remain obstinately shut and the windows
tantalisingly half open. The cars with missing

doors and the vans with tireless wheels are salvaged from the breaker's yard of charity and recycled in miniature eye-level docudramas. Cardboard castles and cut-out houses from cereal boxes provide the random décor for the more sophisticated plots but the main function of buildings is eventually to be demolished by Bad Boys on a rampage who drive recklessly into oblivion, their end announced by loud squeals and your own explosive mouth music … *pppkkkowwwwww!*

The change in scale was always troubling. Going from the Matchbox on the floor to the Ford Cortina in the driveway unnerved you. All this space and your hands sticking to the back seat in the sun-warmed interior of this moving, crystal palace. A stranger to the basic principles of locomotion, you were more impressed by what was on the dashboard than what was under the bonnet. As the shock of the new faded, the car became a part of your everyday world, the part of home that went on holidays with you or kept you dry while you stole away from the house for a quiet read or a prolonged sulk. So when the time came to change the car, the excitement at what might arrive home was bound up with a sadness at parting with this other home, this room with views, which had carried you from your fifth year to your eighth year or from your tenth year to your thirteenth year, a whole commonplace book of memory disappearing into the small ads section of the evening newspaper. All you were left with, like an ex-lover's name and telephone number, were letters and numbers. LZV 514. ZX 438. For months afterwards, your eyes tracked the traffic to see if the familiar shape would reappear, but gradually you resigned yourself to the finality of the separation. The excitement then when you or your sisters spotted the registration plate in traffic, the loved one surfacing in a television report, *Mammy, that's our car! That's our number!* The bewilderment at seeing the complete stranger at the steering wheel of the once familiar car, not your father or your mother but some impostor with the wrong clothes and a wig. Life after your death, you vaguely sensed, might be this – somebody else walking around in your shoes and making a left-hand turn with the indicator of your Ford Escort.

Cars and their owners like pets and their masters became indistinguishable. The angular tail fins of the Ford Anglia seemed perfectly adapted to its spruce driver, Miss O'Farrell, the primary-school teacher who was all regularity and clipped severity (our incredulity the day tears stained her carefully made-up face as Bobby Kennedy's death was announced in Firhouse National School on the farmer's daughter's transistor radio). The black Morris Minor that carried the God-fearing Cassidys on their solemn journeys to and from daily mass was of a piece with the dark dress of the clergy and the family's doleful piety. The red Fiat Cinquecento was sure proof that Barbara Clarke's life was one long, faintly sulphurous carnival going

from the Television Club to Barbarella's.
Sometimes, there was a mismatch. You could never
understand how the austere height of the police
detective down the road could fold up into his
Mini and you were always half expecting his head
to burst through the roof in a cartoon explosion of
metal and hair. Or, as the result of the labyrinthine
negotiations that were the prelude to the purchase
of any car in the house – budgets and expectations
eternally at war – your parents inexplicably (for the
neighbours) decided to buy a second-hand
Triumph Herald. The rakish, playboy abandon of
the Herald appeared to be decidedly at odds with
the solid pretensions of this young, suburban
family. The racy reputation of the car was partly
linked to its recreating the sensations of the early
days of automobile racing. There was indeed no
suspension to speak of and being carried along in
the car was like being pulled along at speed on a
skateboard. As budgets dwindled, so too did the
fortunes of the Templeogue Rake. Driven beyond
the limits of its natural life, and with no money to
carry out essential repairs, the Herald came to a
final halt in front of the garage doors. Here began
the car's ignominious descent to amusement arcade
where for hours on end you and your friends put
the rust-soiled automobile through its imaginary
paces, fleeing delinquent cops and predatory drug
barons. The Herald, an old roué now forced to
humour his nephews at weddings and christenings,
was finally rescued from its irretrievably fallen state

when the Corporation sent out a truck with a crane to take the car away to be scrapped. As the fairground pincers seized the Herald and it swung unsteadily away over the hedges into the waiting truck, the great wreck in the sky seemed like a warning to anyone who would own a car beyond their station.

Ferried for years in the cars of others – parents, friends, lovers – your turn comes to take the wheel and master the dark arts of the mechanically propelled vehicle. The Schools of Motoring become Schools of Life as the instructors come and go in quiet despair. There had been an earlier abortive lesson with your mother in your teens but shouted instructions (*The other pedal, you blitherin' eejit! The other!*) as the Austin Hillman reversed at speed into a ditch did little to restore an already fragile domestic peace. The car was eventually retrieved but any hope of future instruction was ruled out by visible dents to the rear bumper, the modern stigmata of the incompetent. Driving lessons begin in North Dublin where an amiable ne'er-do-well after explaining the basic routines orders you to drive him to distant suburbs to pick up other customers, all female. The instructor then takes up position in the back seat and there is great shared hilarity with the paramour as you juggle with gears and pedals and mirrors in your pilgrim's progress from Kinsealy to Glasnevin. High points are when the car stalls in mid-traffic or is wracked by convulsive coughing as you engage the clutch and accelerator in complicated dance steps. The ribald laughter from the back makes you feel uncomfortably like the dim-witted younger brother bringing an older sibling to the cinema with his knowing girlfriend, the heavy petting and muffled giggles drowned out by the mechanical howl of wrong gear changes. You decide that being a chauffeur is not your calling and try another phone number, the recommendation of a friend. You are slightly nonplussed when the instructor turns up at the house and stares at your shoes. "They're not leather," he eventually says. When you look blank, he adds, "They don't have leather soles. You won't feel the pedals." As you hoofed it over to the instructor's car in your Doc Martens, you sensed that all would not be well. Crawling through the hushed suburbs of South Dublin for three weeks became a journey into the distilled prejudice of Middle Ireland. Spanish students (*Beep that horn – they wouldn't carry on like that at home*), Northerners (*Mind that car, did you not see the reg? There might be a bomb in it*) and Women (*Dawdling in the middle of the road, should never be let near a car*) were the most audible targets of his complex and multiple hatreds but, when you too took your place in the gallery of the criminally suspect and socially undesirable for challenging his bitter voice-over between two sweaty gear manoeuvres, the relationship was set to end and end it did with a rancorous phone call. Charlie, an ex-rally driver, was blessedly free of both erotic ambition and social venom and seemed to view lessons as primarily an opportunity for a quiet smoke in tolerable company. He was the motoring equivalent of Stoicism in the philosophy of the ancients and appeared to believe that it was Fate rather than anything he might do or say that would persuade the authorities to give you a full driving licence.

The Test Centres have the peculiar emptiness of Garda stations, the bleak collage of official notices and the scattering of chairs. In the waiting-room, there is the morose silence of the condemned, waiting for their names to be read out in an official bark as if outside the firing party was already taking up position. Lads with tight haircuts sit, bent over, no longer so cocksure, and make their fingers crack with a mixture of nerves and impatience. Young girls with heat rashes make nervous conversation with their mothers as their fingers mangle the appointment letter. The more mature applicants sometimes engage in a kind of strained banter (*Third time lucky, what!*), a scaffold humour cut short by the disappearance of one of their own into the test room, only to be led out again, silent, a blanched patriot walking into the mocking sunlight of the test-centre car-park. The driving testers themselves have the style sense of Special Branch detectives, all sports jackets and slacks in winter, shirts and ties in summer. You have done a number of these tests over your driving life so you have had time to observe the change of dress and the passage of the seasons. Inside the car, the driving testers begin their slightly weary recitation of the *dos* and *don'ts* (*Drive as you normally would* ...) like veteran flight attendants repeating safety routines on their umpteenth transatlantic crossing. As the test progresses, the tester plays a game of *x*s and *o*s on a form that is carefully shielded from view. The unnerving presence of a mute stranger in the passenger seat does little to improve your humour as you realise you have crossed your arms while turning the steering wheel or coasted to a halt before a stop light. You sense you are being prepared for the drop.

The conviction hardens when you are asked to do a reverse turn. This peculiar manoeuvre involves driving backwards around a corner while remaining close to the kerb. When you have cleared your throat, taken a deep breath and given the knob of the gear stick a vigorous if useless shake, you then turn your head around to realise in growing horror that the car is moving all right but that the kerb has disappeared from view and that the only thing in your immediate line of vision is a box of Kleenex tissues. After a cautious crawl backwards, a vigorous twisting of the steering wheel to the left is dictated by a hunch that the bend is near. It is. Too near. The loud whine of the wheels against the kerb rises like so many cries of the damned. The shirt you are wearing is suddenly too tight and the damp continents of sweat spread,

joining together to form a large, primeval land
mass of perspiration. The volume outside is turned
up as the tyres spin helplessly against the
unyielding concrete. Then the car tilts on its axis
and goes into orbit as the vehicle mounts the grassy
verge. In the sudden intimacy of zero gravity, the
instructor leans into you, tight grip on the
clipboard, and in reply to your strangled question
(*Will I go on?*) he bravely answers "Yes," as a
lifetime of testing flashes by in an instant and he
wonders if he will ever return to the base station in
Rathgar and the prosaic camaraderie of tea with
the other pioneers of road travel. For you, finally,
as the car doors swing shut, there is the long, silent
trek back to the open confessional of the centre
where the tester with the well-practised tact of the
therapist or the hospital consultant breaks the bad
news. *I am sorry to say but on this occasion …*

Maybe the inability to persuade the
Department of the Environment that you are a fit
person to hold a licence is genetic, like eye-colour
or a fondness for the drop. On your mother's fifth
attempt at the test, when the chart in the passenger
seat came to resemble a map of the Allied landings
in Normandy, she stopped the car at a crossroads
and ordered the hapless tester out. Your sympathy
was with the tester years later when an elderly nun,
a friend of the family, suddenly broke into loud
and furious prayer on a slip road of a major dual
carriageway. Turning left instead of turning right
brought your mother and her stricken confessor

face to face with the angry headlights of startled commuters who saw the four horsepower of the apocalypse bearing down on them, with your mother taking moments to realise the connection between the outburst of noisy piety and the anguished hooting of horns.

The rhythmical clicking of the headlights being switched from full to dipped as the car moves through the night. We are all bundled up in the back seat, refugees from the annual holiday streaming back into the city, the sleeping faces fleetingly illuminated by the passing lights of pubs and service stations. The decoding of registration plates (*What county is ZD from?*), the model spotting (*A Ford Consul! I saw it first!*) and the carefully paced incantation of "I spy with my little eye …" are now an eternity away in that remote morning when the suitcases were piled into the boot with the rag-tag army of plastic bags and stray wellington boots. You remember seeing a German film about a left-handed woman where the first ten minutes is a conversation in a darkened car, the speakers picked out by the random lights of the roadside, and how you were suddenly brought back to that childhood darkroom of exhaustion, the needle of the speedometer flickering in the gusts of acceleration. The shock of the stillness when the engine died away in the driveway and how your ears carried the noise of the motor to your cold pillow, the phantom rocking of the car transporting you to sleep.

Each life then a kind of car showroom, not enclosing the gleaming expanse of buffed sheet metal but the ramshackle convoy of our days. The great, dark lumbering Hillman, a fifties behemoth that brought you from Holles Street to home and a convenient prop for the rare snaps of your parents and the plump pudding that was you in the driveway of their new house. Next in line is the grey Cortina which in a magical intermission brings you and your sisters through the Wicklow mountains to rest at tea-rooms with picture-story pots of tea and crumbling scones. Following close behind is the black Mercedes carrying the bereaved family, the car pausing briefly in front of the driveway where, three nights earlier, an ambulance had pulled out in a fury of lights and sirens. Marking time noisily is the fawn-coloured Austin A40, your mother playing footsy with the accelerator to stop the engine from cutting out and ignoring furious hand signals and shouted expletives as she squints into the middle distance on the fast lane of the dual carriageway. Struggling bravely in the line behind is the Renault 4 with the gear stick that juts out of the dashboard and the windows in two halves where one half slides open like the information hatch in a public hospital. Your arm is stuck out of the open half clutching an onion which you use to rub the windscreen with frantic, vigorous swipes. You are on the motorway to Le Havre and the rain has run riot. The windscreen-wiper motor surrenders and a half-remembered story about the absorbent properties of onion has your arm doubling up as a wiper in a landscape of blurred tears and rainfall. As your forearm moves up and down, the fast cars swirl by in a spray of water and silent laughter. Closing in on the Renault is the vintage Citroën convertible and it is a day of celebration, the wing mirrors sporting white ribbons and your father-in-law coaxing the engine up the village hill. In the black and white photograph that turns up in the album, your hand is raised and you look like a visiting potentate about to fall victim minutes later to an assassin's bullet. In the wing mirror of the convertible a car can be picked out, a Volkswagen Golf, and the couple inside are going through the tense courtroom drama of map reading, the angry volley of prosecution questions about manip-ulation of fact (*How you can you say we're near the sea, the car has been going up for the last half hour?*), evidence ignored (*Did you not see the signs two miles back?*) and reckless disregard for the truth (*What do you mean we're nearly there, there isn't a house in sight?*) and the defence angrily refuting the charges amidst allegations of lousy signposting, bad maps and poor directions. And so the cavalcade goes on, wending its way through the narrow streets of recollection, from the scratched Ford Capri on your bedroom floor to the dark Mercs – that sudden luxury of the bereaved – conveying relatives on the last journey by road. As if all licence is indeed provisional.

PLAYTHINGS

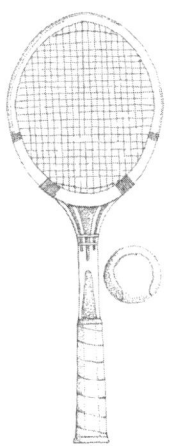

THE PROPELLER DRONE IN THE HIGH, BLUE SKY –
the wood pigeon of the suburbs – the long,
prolonged mechanical call of the light aircraft, a
signal that summer had come to stay. And if Icarus
had looked down from his twin-engined crucifix
inching its way through the heavens what would
he have seen? The slight wobble and then the
furious piston push of little legs driving the two-
wheeled bicycle down the Long Path through the
Back Field. The path that seemed immense from
the splayed saddle of the tricycle, now shortened
with each passing year and the acceleration of the
ever more confident limbs, more runway than
footpath as your voice crackled imaginary
commands to air traffic control and squadron
leaders. On the way, there is an obstacle, Kevin on
his two-wheeler with stabilisers. The curious
drunken motion of the bicycle with stabilisers,
lurching more violently from side to side as it picks

up speed and the impression of pedalling through mud as you press more determinedly and desperately on the reluctant pedals. The main decision in the stabiliser-free delirium of motion is whether to break sharply at the end of the path or to leave it and hit the road with a bone-shaking smack and a nut-loosening thud.

In the house that overlooks the Back Field, the architect is busy. You are assembling Homes for Heroines. Your father has brought home packing materials from work, spongy L-shaped blocks in mottled colours that are the building blocks for your mini-estate. The truth is that you are more interested in interiors than exteriors and in a dreamy monologue of domesticity you begin to furnish the front room with Senior Service cigarette boxes doubling up as settees and an empty box of Friendly matches beams the single channel into the black-and-white world of the expectant show house. The double beds are two empty boxes of twenty cigarettes carefully aligned one after the other while the single beds are tobacco-scented boxes of ten tucked neatly into alcoves. Three or four matchboxes piled high make up the chests of drawers in this Scandinavian utopia of frugality and clean lines. Raffia paper seconded from school projects provides all the soft furnishings, the curtains, table-cloths and rugs. You then go in search of prospective tenants. There they are, up against the wall, the thin, lithe dolls and the pink, swollen, flesh-coloured ones that always seem to have one eye permanently closed and an arm raised behind their heads like a karate chop forever frozen in infancy. Some dolls standing up, not quite straight but at an angle, as if they have had one Martini too many at the gala opening and are hoping the taxi will take them home soon. Others are slouched on the carpet, a snapped elastic meaning that they must forever view the world from down below. These dolls are dressed in the motley materials of household waste, their outfits changed each day on the unpredictable catwalk of childhood whims. They have their two-dimensional cousins, the cut-out dolls from *Mandy* and *Bunty*, *Jackie*'s sensible younger sisters. These dolls all look like air hostesses from 1960s advertisements, their printed smiles hovering over the mauve twin set and the lilac-blue, floral-patterned dress. The homeless extras are carted off in a bundle to their new house – more film set than home – where they are carefully arranged on the existing furniture and made to engage in the polite conversation provided by your sonorous, afternoon-long voiceover (*Anybody for tea?*).

In another room in the house, the soundtrack of engines and explosions and helicopter blades threshing through the weekday afternoons after school. The urgent dialogue. The commands barked at your plastic midgets assembled on the mantelpiece ledge, preparing to abseil to a certain death (*aaaaarrrggghhhh!*) on the shiny tiles below. The planes wheeling around these miniature

dramas of destruction are the makeshift products
of the windfalls of birthdays and Christmases. The
kits with their carefully numbered fragments held
together in plastic rectangles that once emptied
looked like pointless mazes. And the instructions
that always let you down at a crucial moment so
that no matter how many times you scanned the
illustration the piece in your hand was doomed to
remain part of a jigsaw puzzle you could not
complete. The transfers with the wing markings
that slid effortlessly off the paper in the shallow
water of the saucer but which suddenly reveal
ominous fault lines as they are eased off your
thumb onto the wings gorged with adhesive. Then
the sticky fingers of age. Old galleys and biplanes
had impossibly small parts so that the shrunken
spars and struts drowned in the pools of glue on
the tips of your fingers and what followed was a
stringy tug-of-war as you tried to free the part from
your finger while leaving it on the underside of a
wing or in its rightful place behind the mizzen sail.
The larger the box, the older the weapon of
destruction and the more dramatic the illustration
on the cover, the greater was the certainty that
what lay within was a construction nightmare. You
admired but were nonplussed by friends who had
the models, fussily camouflaged and complete with
serial numbers, mounted on stands. The frozen
museum of adult admiration (*Have you seen Jim's
Hercules? He's been at it for weeks*) appeared too
precious for the exhilarating anarchy of play,

missing propellers, smashed vertical tails or
dismembered rudders making up the true
inventory of fun. The boxes crowded with the
booty of birthday parties and School Fairs, the
carefully hoarded capital of childhood, added to by
the cautious barter of the playground – marbles,
water pistols and model cars changing hands as the
school bell announced the beginning of break and
the opening of trading. How you slowly,
imperceptibly shift out of the passions for
playthings until you no longer add but subtract as
the collection is broken up and dispersed to
visiting cousins and neighbours' children in for the
afternoon.

In and out went Bobby Bluebells and, as the
rope swung round and round, we were urged, as
we had been for years, to vote, vote, vote for De
Valera and we asked knowingly what Queenie-I-O
had the ball, but already television was drawing up
the calendar of our playing year. This was the
decorous age of television where your father would
switch off the lights and draw the curtains and you
settled down in a cinematographic darkness of
expectation. Programmes for a few hours a day
occupied carefully arranged slots between the
geometrical abstractions of the test cards and the
billowing flag of the national anthem. The
unpredictable endings of sports put these tightly
choreographed schedules under pressure but then
cable cleared the rooftops and sports were never
more than a remote-control button press away.

June was the month of School Holidays and Wimbledon and suddenly the roads were filled with the old rackets and balls that bounced oddly as you and your friends lived out the drama of that final set and tie-breaker on the fissured concrete of the road through the estate. The net was an agreed fiction and random lines and cracks were the court's haphazard markings. As the summer slipped through your fingers and school rounded you up for autumn, the first and third Sunday in September were like the last high holidays, colourful *post scripta* to a year that was ending not beginning. And so Micheál Ó Muircheartaigh and Mícheál Ó hÉithir drove you out to the Back Field where you swung the hurley and soloed with the ball in laborious parody of the Bannermen and the Lilywhites and the Dubs and the Men from the Kingdom and the Rebel County who had sprinted across the living-room screen hours earlier. The high drama of your ragged, breathless commentary (*It's a high, high lobbing ball!*) compensating for the poorly aimed smack of the tennis ball with a bandaged hurley.

Every four years came the global marathon of the World Cup and the Olympic games. Spanish students were recruited to perform in our raucous approximation of international football and, as the scented gazelles sped by you (that strange, unmanly foreign obsession with smelling good), you dimly sensed that the nation's footballing future was not in safe feet and much remained to

be done. The Olympic games were to transform
the fortunes of household furniture. The high
jump for you and your sisters involved placing one
end of a sweeping brush on the arm-rest of an
armchair and the other on the arm-rest of the
couch. Books at both ends could serve to raise the
brush and the handle was cleared by taking a short,
furious run from the dining-room through the
open sliding doors, jumping over the handle and
landing in the exhausted cushions of the red and
black couch. The couch absorbed the shock of
these elfin furies catapulted into the air until one
day a loud crack from the splintered base indicated
it could take no more. For years afterwards, the
broken plank jutting out from the underside of the
couch made it look like a ship perpetually in
distress, perpetually about to go under in an uproar
of shivering timbers.

Your nib falls into the tiny blue-stained crucible
of the school desk, the meniscus of light on the
surface of the ink shivering in the crowded
classroom. So many of us corralled in the dead heat
of late spring and the Carmelite bell from the
nearby monastery sounding through the ragged
polyphony of our ten-times tables. The bee-swarm
into the yard at break-time and the explosion of
play. The great Roman circus of your
entertainments as the generation nursed on
cathode tubes prepared for the excitement of the
Colosseum. Two of you held hands while the other,
a Ben Hur in shorts, held on to the jumpers of the

knobbly kneed horses and with a fierce tug and angry yell drove them forward into a dumb-play of combat. The exhilaration as the chariot teams circled the walled field, wheeling and turning in a world before insurance. More medieval pastimes were provided by piggyback jousts where the rider attempted to unhorse his opponent while his unfortunate steed wheeled and sweated in a world of wet wool and breathless curses. In the charge of the heavy brigade, knock-kneed nine-year-olds carried howling knights from fifth class, the hollering warriors urging on their stumbling chargers with vigorous stirrup kicks of their Clark's sandals. And from the yard, the steady clap, clapping of the hands on thighs and palms. And twist, twist, twist like this. The elaborate daisy chain of elastics cutting into your legs and the taut figures of eight as the elastics rise higher and higher and your breathing grows louder as you soar higher with each new jump. The corners of the yard have been staked out by the players of the glass-bead game. The shrunken pinball world of marbles. The swag bags eased out of trouser pockets or bags or coats with their carefully audited contents. Inside the bags, the small, workaday glass marbles and the élite division of golliers and steelyers. The golliers were the large marbles, the giant planets in the system, and the steelyers, science-fiction pellets of hard steel, ball bearings that had wandered into a new career. Then, in a class apart, the milky marbles, like porcelain beads, the oriental

fastidiousness of their decoration seeming all too
precious for the rough-and-tumble scuffle of the
schoolyard. Rolling a marble along the bowling
alley of the pen rest on your school desk you
wondered how the bright yellow or blue leaf came
to be trapped in the perfect ball of glass. These
ordinary miracles of industrial production were
easily sacrificed but for the steelyers and the golliers
the stakes were much higher – the tense intake of
breath and your heart tracking Damian's gollier as
it homed in unerringly on your prize possession,
now about to disappear into Damian's monstrously
inflated plastic pouch. Walking home on the good
days, your pockets bulged with the hard, molecular
mass of the expanding marble collection. In
chemistry class at secondary school, as your
attention wandered from the molecular geography
of actinium, iridium and nitrogen, the whole
atomic world began to dissolve into a microcosmic
game of marbles, the atomic and sub-atomic
particles rolling and striking off each other in some
fast and furious play of the elements.

The mahogany sheen of the chestnut. The
crafted smoothness of its surface as you broke open
the unpromising outer shell, a pale landmine of the
Great War, the prickly fruits clustering in the
undergrowth of autumn. The stillness of the
moment as you stood there, the pierced conker
hanging helplessly from the loose string, and next
the formidable crack as the wrecking ball of Joe
Farrelly's chestnut split your bloated atom into a

scattering of fragments. When later you went to Paris and winter turned into the smell of roast chestnuts in the rain (*Chauds! Chauds! Les marrons! Chauds! Chauds!*) you kept expecting to see chestnuts roasting on the pavement barbecue shatter into the smithereens of lost conkers.

The semi-organized chaos of primary gives way to the barracks life of secondary. You are assigned to the 14D rugby team, a collection of the incompetent and the insubordinate who are bussed with a reluctant priest to the scattered playing fields of professional Ireland. The 14Ds, this team of conscripts, a Black-and-Tan rabble of sporting misfits, would be emptied out of the old coaches under the close supervision of the clerical orderly who had been given the posting as some form of punishment for sins in this life or the previous. You wore a Shamrock Rovers' jersey and flanking you in the front row was your friend wearing the colours of the Kerry County football team and the new boy from Portlaoise in an Arsenal top. An understandable fear of poor results and an obscure shame at the school having spawned this monstrous offspring meant that you were not encouraged to wear the school jersey and allowed instead the harlequin costumes of the sporting damned. The rhythmical crunch of studs on tarmac announced the beginning of these ill-matched encounters where the rain fell with increased momentum as the weary PE teacher noted the score for the other teams rising through

the double digits. The scrums were a trip down the
ladder of evolution where grunts, heavy breathing
and attempts to bite any fleshy appendage within
reach were fair game. As the Hooped Hooker, you
were suppose to get the ball out of this primeval
darkness but physical survival became the priority
as the obstreperous crab turned and buckled. As
delivery was unerringly poor, you were transferred
from the scrum to the full-back position. Through
some sadistic shuffling of match fixtures, your
team almost invariably found itself playing against
B teams from the mercilessly ambitious hatcheries
of the country's sons of substance, presumably the
matches a kind of blooding for novices before they
went off in pursuit of more praiseworthy prey.
Standing there in a Flanders' field of mud and grey
cloud, your role was to stop the advancing front
line as it breached the defences of Kerry and
Arsenal and raced towards you, the last sodden
sentry defending a city from which all the other
inhabitants, in the shape of two full-backs kitted
out in the colours of the Kilkenny hurling team,
had fled. More like Custer down on his luck than
Canute armed with conviction, you waited for the
inevitable. If you were lucky, you were briskly
outflanked and all you felt was the articulated
tailwind of some teenage Titan as he rushed by to
expire in an orgasmic roar on the touchline behind
you. In less fortunate moments, a testosterone
torpedo would lock in on you and, slicing through
the grass with inhuman speed, it would hit its

target with appreciative roars from the other frontmen of destiny, as you crumpled to the ground with the distant convoys from Kilkenny looking on bemused. As you were herded back on to the bus after the hard labour of play, the attendant priest, shifting momentarily from warder to counsellor, would exclaim, "Could've been worse, boys. Could've been worse!"

The streets and roads are colour-coded for the prosperity of games. The first time you walk down Ailesbury Road it is not the leafy embassies and the sophisticated intercoms of the richly paranoid that impress you but the fact of walking on the most expensive square on the Monopoly board. In the city as board-walk, on Pembroke Road you half expect the sturdy residences of the well-got to dissolve into the plastic green prototypes of Monopoly hotels, as apprentice developers trade their counterfeit cash for the lucrative debt trap of rent. On Landsdowne Road, the stadium is empty and you wonder if somebody will roll up that luminous green cloth spread at your feet and take it away for a match of Subbuteo, the forefingers poised to strike the figurines on their giant, handless coffee cups. The ramshackle, plastic glue surgery that prolongs the lives of these players, crushed by the kneeling knees and misplaced forearms of the restless Leviathans towering over the cloth of green. In Stephen's Green, tiny feet are stretching towards the sky as you get pushed higher and higher, tight fists gripping the rope until you get older and bolder and on a lonely impulse of delight leave the swinging seat and go into free fall. In Bushy Park, the see-saws creak through the weekends, the seats rising and falling with two miniature figures at both ends, accelerating away from the ground with forceful grasshopper thrusts of their tiny limbs. In St Anne's Park, two girls on a bench begin the long novitiate of card playing, from the savage initiation of Snap to the permed sociability of the Bridge circle.

And the plane up above is coming down now, dropping through the fading light of the bedroom, the instruction to air traffic control fading into darkness and the tired arm of the infant aviator on the bed stretching finally towards sleep.

PAGES

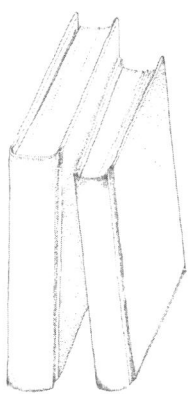

LIGHT HAS PICKED YOU OUT ON THE PILLOW. IT IS morning and the clock tells you you should be up, but this is the childhood holiday of sickness – comics, tea at different hours of the day, Lucozade and the sudden, anxious solicitude of your parents. Your mother tousling her hair and slowing time to read stories and embrace you in the memories of her own past – John Hinde shots of pony and trap and Hops in the National and green buses by the Pillar. You saw the colours fade as the stories grew old but as a child you loved those islands of complicity, days off that seemed flushed with the light of shared, easeful confidence. Later, in the last two years of secondary school, you dreaded staying at home, under siege from your mother's worried questions about exams and girlfriends and drink and whatever else was jamming the lines of radio phone-ins. You wished your mother was at work all day so that you could smoke your cigarettes in peace rather than having to leave the window open and invest in the poisonous aroma of supermarket air-freshener. Now you are propped on the pillows again but it is the weekend and the particular privilege of the healthy is this confinement to bed as daylight clamours for action. On the floor is no longer the untidy spill of dog-eared copies of *Tiger and Jag* or the *Beano* but the layered remains of the week's newspapers and the rising columns of books half-begun and half-finished. You think of the veteran anthropologist you once saw on television cowering under the tidal wave of books in his Parisian apartment and how he ruefully confessed that he had spent his life battling against the books that invaded every square inch of his living quarters. He had lost. The books had won.

Eyeing the victory columns beside your bed, you know the habit starts early. The Ladybird Readers in High Babies, with Peter and Jane swinging into the sun as you climb the ladder of literacy. The scattered islands of words in the beginning and then the black print swarming across the pages as you go from 2A to 4B to 6A to 12C. The day-glo world of Home Counties suburbia lit up by the eternal sunlight of the illustrator frames the uncomplicated lives of Peter and Jane as their smiling parents take up their allotted positions by the cooker and in the armchair. Then there is the enigma code of the front page of the daily newspaper that gives up its secrets one day in the kitchen so that learning and

life collide and you can begin to parse your parents'
conversations. As words take on this new power,
the hunger grows. The Secret Seven are first up
until they are dismissed by the withering
condescension of the older child as too "babyish"
and you move on to the more demanding fare of
the Famous Five. Their adventures too will fall foul
of the derision of more adventurous readers in your
class and it is the turn of William and Bunter to
bring the highly coloured exotica of English
boarding-schools into your day-school life.
William's harmless truancy was somewhat easier to
follow than the rash of exclamation marks that
charted the torments of the hapless Bunter so that
you could never figure out who was shouting what
at whom at what moment except that Bunter was
the inevitable fall guy in a splatter of punctuation.
The brisk competitiveness of those years, the titles
you had not read like so many football cards you
had not collected or marbles you had not won.

Coins from visitors and the chance windfall of a
note from relatives on an annual visit bankroll the
longing for the artificial paradises of the page. The
books knocked back and the anxious, urgent
scouring for more. The Sale of Work or the Garden
Fête or the School Fair were bright oases of
opportunity in the grey wastes of the school year. The
eyes scanning the boxes for the orange or light green
spines of Penguin Modern Classics, the off-white
spines of Picador or the black elegance of Penguin
Classics. Or raking through jumbled piles of Harold

Robbins and James Hadley Chase and Barbara Cartland for some unnoticed and under-priced catch. The slightly sweaty arithmetic of small budgets, wondering if you could go the extra ten pence for the Orwell or the dog-eared Bulgakov. Going away and coming back again and rereading the blurb and the sober proprieties of the biographical sketches (*James Oakes lives in London with his wife and two children*), and inspecting the scratched marginalia (*GOOD!*) and the smudged signature of the former owner (*John O'Hagan, Blackrock, 12th June 195—*) and feeling light-headed at the prospect of pennilessness that would hinge on your decision. As you trailed home, you asked yourself if you should have bought the other book and begin to dream of a life where wealth would abolish the tyranny of choice.

It was a dangerous business, reading. As you shifted uncomfortably in the armchair under the irritated poking of the vacuum cleaner you were reminded that all these books would *ruin your eyes*. Sprawled on beds, stretched out on couches or slumped over the kitchen table in the wind-shaded warmth of a late summer's afternoon you were told that *You should be out playing with the other children and getting some exercise*. Here you developed that defective other sense of the reader, selective deafness. Calls to meals, summonses to bed, would be carefully filtered out from your attention as you determined to finish your chapter. As the cries became more insistent, you quickly leafed through the pages to see how many were left and afterwards held your breath as your eyes raced ahead and you hoped for a phone call or a surprise visit from a neighbour to stall the explosion of irritation and the confiscation of all reading matter. As days merged into nights, the threat of the printed word did not abate. Nightly patrols from your parents' bedroom would check for any rogue lights, any sign that the reading blackout had not been observed and that you were there in your bedroom gorging yourself on words while your poor eyes withered from exhaustion. The tunnelling into light began under the blankets with torches and the ever-present anxiety was that coming up for air would reveal the stolen beams and lead to an immediate routing of your position. Later, as the child mutates into the adult, you wonder how long you can leave the bedside lamp shine into the face of your sleeping partner as you try to dodge the white nights of insomnia by reading a few pages. You ask yourself as you lie later in the dark how many more there are in the city who need to be coaxed into sleep by the lullaby of sentences.

Your first visits to a public library were alarming, like an addict being confronted with the confusing riches of a pharmacy. When you were duly handed the blue tickets into which the slips of paper from the Chosen Books would be inserted, your only thought was of overdose. How would you cope with the vastness of what was on offer in Rathmines Public Library? History, Fiction, Biography, Fine

Arts, the varieties of passion were endless and the books themselves appeared more substantial, cosseted in the transparent armour of heavy-duty plastic, popular books riddled with the puncture holes of use. Inside the libraries the ecclesiastical silence would be broken only by the confessional whispering at the issue desk and the occasional elderly reader struggling with volume levels that startled everyone except himself (*I LIKE JOHN D. SHERIDAN VERY MUCH*). A hidden terror of this nursery of earthly delights was the cost of amnesia. Failure to return books on time set a punitive arithmetic of pennies in motion so that you were constantly fearful of ending up as the Rip van Winkle of the book-room, waking one morning to find that days of indebtedness had stretched into weeks and months and years and that you were in hock to the library for the rest of your remorseful life. So opening library books, your eyes would nervously scan the date stamp until the studied indifference of your late teens set in and only the formal threat of final notices reminding you that books were overdue would force you up on a bicycle and off to the issue desk where you reluctantly parted with your coins in the debtor's poor box.

There were other books, of course, schoolbooks that no amount of colour or illustration could camouflage as to their true purpose. The more dutiful pupils covered them in brown paper with their names stencilled on the front or encased them in thin sheets of plastic, their name flush with the crisp edge of the title page. Occasionally, household decoration would lend grandeur to the regimented ordinariness of the textbook, the books covered in stiff pink or blue wallpaper with embossed motifs. You made attempts with plastic shopping bags to emulate the comely bindings of your fussier classmates but the Sellotape never seemed to land where it should and when you closed the book the plastic billowed in all the wrong places. The brown-bag covers that you tried gradually mutated into doodle pads as the interminable liturgy of the schoolday sent the mind off in pursuit of ink-splotched cubes and pop-art squiggles. In the School Fairs you would come across the textbooks that were no longer textbooks and they appeared like the Lost Ones, books that were to be forever abandoned to the *auto-da-fé* of the unloved and the unread. The Books that were No Longer on the Course, so many toppled statues to be left in the wrecker's yard of the bargain basement as all hailed the regime change of the new syllabus.

It began in outhouses. He said you could come anytime you wanted to look through the books. The local parish priest had once served as a chaplain in a university. Raising funds to make the inevitable repairs to the roof of the church had led to desperate measures such as the recycling of printed matter and this included books. Trucks brought consignments of the condemned from the university library, books that were no longer

needed or no longer read or were unsolicited gifts
from unwanted benefactors – the ex-libris in the
inside cover akin to a dinner invitation politely
refused. Like a minor government official faced
with brutal orders from a foreign force of
occupation, the former chaplain delayed,
prevaricated, made up elaborate stories as to why
the books had not left for their final destination.
His unlikely saviour was a sixteen-year-old hungry
for print with a dream of a library he could picture
but not afford. So for weeks on end you trudged
across the Dodder, bringing the random cast-offs
of a great library to safety in an assortment of
banana boxes and black bin bags. As the outhouses
emptied, your bedroom was invaded by the sweet
smell of faded paper and the dust storms of books
finally released from the captivity of indifference.
Here on two cut-price bookshelves from a
furniture discount store and shelves put together
from poorly sanded planks and filched bricks, you
assembled your scale-model Library, a bubble of
stillness in the growing estate.

You are almost sixteen and it is the Summer of
War and Peace. There is a heat wave and no rain
falls to stop play, like August 1914 or some hazily
remembered pause before the light goes out of the
world and everything becomes more complicated.
You are lying on your side on a couch of mowed
grass, self-importantly cradling Tolstoy from the
sun as you pick your way through the missal-thin
pages of the Heinemann edition. In the almost

audible silence of the heat, you chew grass and tease ants while your mind toils after Napoleon in the winter snows. These are the summers of the heroic reading binges of adolescence. In the gardens, on the buses, in the trains, on the beach, on the couches, in the beds, there is the prolonged ache of print, light darkening to night as the thumb turns and turns and turns. And in the awkward age, the false modesty of the reader as, on the buses and in the pubs, you shield the title of the book you are reading, fearing ridicule or the wry smile of derision but twist your neck and raise your head to make sure that you find out what everybody else is reading. In the blind date of readership, there is always the hope that a shared title or a preferred author might release you from the Coventry of shyness and that a life of bookish intimacy might start with the glimpsed page.

The peculiar privilege of making a living out of books, like an obscure favour handed down by the gods or a royal privilege dispensed in a lightning flash of charity. He always stood erect, his lank, shoulder-length black hair highlighting the deathly pale skin of a Late Romantic who had wandered into the world of bookselling. His cut-glass accent that articulated ever so carefully the titles of the philosophy books that were mainly in his care and his intense fixing of the middle distance as you teased out your halting query gave him the majesty and authority of a prophet, as if he had been called upon to consult the oracle of the ancients and not simply find you a book that had strayed on to a college reading list. His unwillingness to smile and inevitable dark dress marked him out as one of the Bohemian Elect, an angel of the word come to hand out primers to ordinary mortals. Nearby was the first-floor bookshop beside the market with the second-hand records, Indian beads and tie-dye T-shirts. Here were the Radical Books and the photocopied pamphlets and closely printed newspapers and magazines from the endlessly gradated varieties of the Far Left. The wooden floors and shelving and the quiet austerity of the owners seemed of a piece with the dark intensity of *émigré* intellectuals dreaming of uprisings and radiant dawns. This was the political wing of the amiable patchouli-doused anarchy of the market below. Two streets further down, the bookshop at the end of the poster-lined tunnel. This is the last salon for the city's boulevardiers where, between the remaindered classics and the Sellotaped LP sleeves, the readers-as-talkers let themselves go in an endless riverrun of conversation, the envy of their employed friends. And so where books congregate, they gather, the lost children of Gutenberg's revolution.

You see them, victims of cruel indecision. Dragging bags, lugging cases, stumbling down stairs, wobbling on bikes, breathless in airports, exhausted in stations – the long distance book porters. Rucksacks stuffed with one change of socks, two pairs of underwear and fifteen books. An hour before you leave, the whirlwind tour though the shelves and suddenly it is the beggar throng of titles around the departing reader, all clamouring for Your Honour's attention. The fictional and the factual, the long and the short, the domestic and the foreign call out for the pittance of your attention, not wanting to be left behind in the empty city. Your mind runs through an anxious arithmetic of desire and possibility, the nervous helicopter pilot waiting to leave the compound as the rebels close in, and you wonder how much room is left and how many titles can be rescued from the unenviable fate of the unread. So, of course, you bring too many. As you labour down the interminable tunnels of international airports, not a trolley in sight, and your right shoulder sagging under the weight of the print cargo, your thoughts wander to the Via Dolorosa and you wonder do all readers carry the cross of their addiction through life. When you get to the destination, what you have brought no longer seems enough and there is the allure of other bookshops, other titles. By the time you are homeward bound, the load has swollen to incorporate these new stowaways so that you increasingly feel like a benighted Egyptian slave hauling some impossibly heavy stone towards a pyramid situated at the gate that all flights to Ireland seem to leave from, the farthest possible one from the check-in desk.

The straps are eased off your sore shoulders and the rucksack slumps against the lower bunk in the hostel. The pressure of the load that lingers reminds you that wherever you go you wear the hair shirt of literacy, the unread books that crowd out the jumpers and the T-shirts. You are in Venice and later that day you see a large painting showing St Augustine in his study, his books carefully arranged to spell out the contents of his mind. The saint, quill raised, turns towards the high window. He has learned of St Jerome's death and yet his face is not crumpling, there are no tyre tracks of tears, no hand raised to remove the salt from cheeks bruised by grief. The features are alert as if he has not heard enough, that there is more to come and what comes after is the slide show of the reader's memory. The hours on the old armchair in the shadow of the swing, slightly dazed by the great draughts of print, or the bicycle inching up to the Pine Forest, the library books bulging precariously on the saddle, or evening on the second floor of the college library, deserted by the coffee-bar talk of daytime and filled with the sudden intimacy of that vast emptiness. The book pages stained with in-flight service and rushed breakfasts. The bereavement of book endings where for days afterwards you wander around looking for the impossible sequel, a jilted lover hoping for a reprieve. The saint puts down his quill, no longer listening for the watchful steps of the parents who are no more, and only then will the books on the shelves in their unending hymns of prose call all readers out to play one final time with Peter and Jane.

THE ICE AGE

IN FITZWILLIAM SQUARE THERE WE WERE, ALL three of us, sitting on deckchairs, ranged in a semi-circle around a Super Ser heater. Drinks in hand, legs stretched lazily towards the dying sun of a gas bottle, we were islanded by the improvised and improbable luxury of the moment. A step up, in a sense, from the electric bar heaters with their orange rods glowing in a thousand furnished rooms. Come On Baby Light My Electric Fire. The bachelor frugality of the one-bar electric fire, a lonely despot in its desire for human closeness. Unless you were hunched over these electric fires you felt nothing so you warmed your finger tips, playing delicate keyboard routines with the rising heat waves, while your neck and your behind courted hypothermia in the gathering darkness. People lit hissing cigarettes from these cylindrical torches and toasted bread grown stiff with the days. The mind toyed with what would happen if a trembling finger went beyond the perimeter fence of the steel grille and tried like a Lilliputian Icarus to make a dash for the blazing heavens of the General Electric Company.

The layers. And layers. The gloves for reading. The thick woollen socks with the funny bed smell for keeping your feet thawed. Your parents' overcoats spread-eagled on the bed as if any form of insulation was welcome in the Irish Ice Age. The vests and the bodices and the cardigans and the jumpers and the coats and the hats and the mufflers and the scarves and somehow it was all no use against the rain and the cold howling down the long avenues of new suburbia. *Put on your coat*, the chant that heralded all our exits and departures. *Put on your coat*, the parting instruction to frisky children and mumbling teenagers. The extra layer, a layer too far, those four words *PUT ON YOUR COAT* summing up the endless harassment of growing up, the incessant artillery fire of repetition as if the opening of a door always triggered some sub-routine of admonition, "You'll get your death of cold!"

The dread of losing these coats and anoraks and scarves and jumpers so that any public place becomes a quantum playground for the absent-minded as homeward-bound they look around in growing panic for that green jumper, that blue jacket, that horrible top. The despondent tread back to the empty hall, the deserted stadium. The careful picking of the mind's way through memory's corridors as you try to piece together your movements over the last three hours. Wondering how long before the missing item is noticed and the official parental inquiry begins. The very young had gloves on a piece of woollen string so that infants on a cold day always looked like rather unhappy puppets without their master. But such artifices were scorned by age and older children had only the fickle companionship of memory, easily distracted by the promise of a good time or the lotus heat of indoor arenas. One day, watching a "claim", a squeaky maul of grimy fisticuffs in a parish hall, you somehow lost your hat, or "helmet" as your mother called it. The helmet was a leatherette cap with a peak, two earflaps and a chinstrap. It was bought in O'Connor's Drapery Shop which also sold "slacks" and "sports jackets". The earflaps were prized by your mother as a redoubtable invention, a charmed way of foxing the Irish winter as you sailed out each morning on your Czech Eska bicycle to go to national school, like some infant Biggles careering into adversity. For you, the helmet was the jester's head-dress, the Fool's Cap and you lived in dread of being spotted wearing it by Dave or Kevin before you were safely out of sight of number 87 and able to stuff the cockscomb's hood into a coat pocket. The disappearance of the unlovely headgear brought only passing relief soon giving way to alarm as you realised that alibis would have to be invented to explain the absence of the peerless protector of children's ears and artful enemy of the head cold. The case was hopeless. The only thing for it was to rustle up funds from the sisters and make your shamefaced way to the dark interior of O'Connors where from a drawer under the glass counter a double was summoned to ease the coins out of your fist. As you cycled home, clutching the crown of thorns in a plastic bag, the cold wind careening in your ears, you yearned for the licence of adulthood, for a world free from the routine indignities of Helmets, Mittens and Bawneen Hats.

The nightly ritual of the hot-water bottle. Ignoring the printed rubber warning you poured the boiling water into the opening that was too narrow, a nervous alchemist expecting any moment to be scalded by the wayward liquid of your experiment. And then holding the hot-water bottle to your chest like a cherished infant, you made the slow journey to the refrigerated container that passed for your bedroom. Or else, too hot to handle, the bottle would be dressed up in an old pyjama top so that it lay briefly on top of the bed, a forlorn torso. Clamped between the two frozen blocks of your feet, the bottle produced an almost sexual thawing into unconsciousness. Then there was the shock of discovery as you woke in the middle of the night, warm but confused, to find this cold rubber jelly hiding out in a corner of your bed. Now the bottle was pariah, shunned by your feet which turned the other way, avoiding any contact with the Cold One, or else in the grumpy half-light of discovery you impatiently kicked it out of the bed until it slid on to the bedroom floor with a satisfying thud.

The clothes dryer in the kitchen, a box-like affair with a light-green lid. Inside were the rails on which you hung the wet clothes. But its main function was as Speaker's Chair, a kind of heated throne for a suburban court where you spoke not so much *ex cathedra* as *ex calore*, the warmth lending *gravitas* to the words spoken from the eternal summer of the dryer. Sitting up there you were invulnerable, a prophet in a Florida of your own making. And when you had to come down from it because of meals or homework or My Turn Now you became a disgruntled Moses, condemned to wander in the ice floes of the unheated house. Sometimes this chilly exile would be cut short by the soft breeze of the fan heater until it too failed to offer any joy, its innards fouled up by the dust storms of carpets and four free-range children.

If cleanliness was a haphazard affair, it was not the soapy sacrament of water that acted as a deterrent but the cold before and after. Through some mischievous paradox the one room in the house where you were likely to find yourself buck naked was also the one without the radiator. The only source of heat was generally one bright-orange bar set high above the door with a cord switch. The effect here was primarily visual. No heat ever reached you from this distant star as you stood trembling, the always-too-small towel imperfectly covering your gooseflesh. The bathroom indeed was every household's version of the parable of Eden. Stretched out in the bath, fully immersed in the blood-warm paradise of properly heated water, your face damp from the rising steam and your thoughts wandering languidly to the evolutionary possibilities of fornication in the primal swamp from which mammals emerged, it seemed that here at last you had reached one of those moments of Ecstasy and Fusion that mystics associated with the Second Coming and our return to an edenic state. But as the water in the gardens of paradise slowly cooled, the steps on the stairs become more audible. The blast of cold air and the summons to Get Up Out Of That Do You Know How Long You Have Been In There were the first intimations of a paradise soon to be lost. Soon you would be standing shivering on the blue bath rug, a naked penitent, head bowed, facing into the punitive ice storm of the landing and the cold embrace of cotton pyjamas.

Households would "get in" central heating but it was rather like the car in the driveway that was taken for a "spin" at the weekend – the mere having of it was all. So the system would be put on for an hour a day or at odd hours of the night; the pleasure being primarily aural – the sound of water sloshing expensively through the pipes – rather than calorific, the house would remain an icebox. *The bit of heat is lovely* was the dismal epitaph on hours spent in the icy wastes of the Good Room, hands cupped grimly around the fading warmth of teacups as Mrs Farrelly priced crude oil with all the dexterity of a commodity broker. Your shoes felt the strain first. Wriggling your toes and arching your sole to coax life back into your numbed extremities. As crude oil prices rose so did the cold, making its way up to your life-support systems. When Mrs Farrelly went to further reduce the trickle of afternoon heat (*Isn't it terribly warm in here?*) you sat on your hands, the undersides of your thighs sensing the polar despair of your frozen paws.

The rooms in houses abandoned to the icy hordes. They remained idle for months, useless in their coldness, hostages to our scale-model version of the Big House. Every house seemed to have this Cold Room of Formality as if acting proper meant a lowering of the temperature. No hastily summoned convector heater or arthritic radiator could ever remove the spell of the Snow Queen so that as you placed knife and fork carefully on the scoured plate, you felt the cold tugging at your elbows like an insistent vagrant. The grate gleamed in frozen idleness. Or the great emptiness was masked by a cheerless fire screen with a picture of an Alsatian dog or a spray of yellow roses. The very words we spoke seemed measured out by depleted reserves of body heat as if the very act of speaking was a mad, shocking sauna dash in the snow.

The great sarcophagus at the bottom of the

stairs, fit for some immortal captain of industry –
the storage heater. The weight of the brown,
oblong box, like some mute ancestor of the
computer, glowing silently in the houses of the
new city. It was the obstacle to avoid in the
downhill slaloms in banana boxes. Sitting into the
box at the top of the stairs, see-sawing on the
precipice of the top step until you finally launched
yourself into full flight by tilting your weight
forward. You had to gauge the size of the gap
between the storage heater and the safe haven of
the hallway. The gap narrowed dramatically as the
world bucked under your knees and only the most
violent swinging would bring you safely past the
juddering proximity of the immovable one.
Sometimes, of course, the stairs reared up in dark,
unpredictable ways and there would be the howl of
impact as box and contents collided with the steel
box at the bottom of the stairs. It was hard to think
of the storage heater as just a heater and not to see
it as a kind of brutish minder, employed by
sceptical parents to control the riotous playfulness
of their offspring. Dents in odd places and missing
screws were the battle scars of this domestic droid.

The roasting spit in the house you shared with
two French students. The bedroom fireplace which
doubled up as the sitting-room fireplace in the
rented house was the sole source of heat so every
fifteen minutes the occupant of the seat nearest the
fireplace would move making way for his nearest
neighbour. You had fifteen minutes in the

bituminous sun before the clock hand nudged you into the face of the advancing cold front. So it went on night after night, we three repeating our routine like giant figures on a town clock endlessly ringing in the quarter hours.

The gap under the door, the ill-fitting window, the high ceiling all inviting the heat to come outside and join the winter festival and leave mere humans to their Arctic dreams. It was difficult to explain why you never made it to nine o'clock lectures in College or rather, in the general conspiracy of frost-bitten hardiness, it sounded a bit lame. *I was too cold.* But it was true. You were. The endless, protracted haggling under the quilt as Duty pleaded with Comfort for a concession. Get up now. Make a dash for it. You will be all right when you are outside. The bus is going in ten minutes. But Duty was dealing with the whining evasiveness of the inveterate smoker and the lapsed alcoholic. Just five minutes. Just five minutes more to enjoy that carefully husbanded warmth, five minutes to savour the precious yield of a night's sleeping, five minutes more lying in the shallow pool of heat surrounded on all sides by sheet ice. Putting an arm over the quilt confirmed the worst. Conditions outside base camp were deteriorating. You thought of the lost worlds of Old French and the Metaphysicals and the Modern Novel and wondered how many more were laid low by the intemperate weather conditions of the Irish bed-sit. Soon despair would set in. The north-west

passage to the landing was impassable. Nothing for
it but to abandon all hopes of rising and lie
huddled under the quilt, compiling a diary of
excuses, like some truant Scott of the Antarctic.
And you would think of your father sitting under
the flickering adventures of Laurel and Hardy in
the winter of 1947, with his engineering textbooks
and army-surplus coat, trying to salvage some
warmth from the cinema and the crowd, his digs in
Ranelagh, a lost Thule, deep in permafrost.

One evening a Young Poet came to visit you in
the Ice Temple. Banished from the loud warmth of
the pub at eleven, it was the usual sprint to the bus
stop and an invitation to finish off the remains of
a six-pack from some previous gathering. Starting
a fire at the late hour was difficult – spluttering
matches, firelighters in economic crumbs that took
fire and immediately guttered like children's
fireworks and coal which remained unmoved by
the entreaty of flame. The Business page of the
newspaper was stretched across the open fireplace
but there was no illumination from behind, no
tiny brown hole bursting into a circle of flame,
only the unpromising smell of coal smoke and
newsprint. Eventually, there appeared a trickle of
light between the black lumps and anxious,
tentative prodding (too vigorous and the fire faded
in a cascade of tiny sparks) produced a narrow
ridge of flame which illuminated the doll's grate.
There was, as we soon realised, light in the fire but
no heat. As you picked your brazen way through

the wreckage of reputations, past and present, you noticed the Poet's nose, becoming bluer and bluer, an Arctic Pinocchio, his whole face an ice mask set against the Siberian hostility of the flat. As the fitful warmth of the beer faded and the night lengthened, the words became rarer and rarer and you would not have been surprised if days later a search party had found you and the Young Poet frozen in an eternity of conversation, hands raised to emphasise that last fatal point, eyelashes matted together in the frozen dawn of a day that never came, a sun that never rose.

The Russian roulette of rented accommodation. The electricity meters with narrow slots for the coins. The inevitable click and the fusing of expletives (*Shit!*) as the room descended into darkness at a crucial moment in your dinner for two. The dish in the oven cooled rapidly and the heat ebbed from the room as you went nervously rifling through pockets and tin boxes for coins to bring the dinner back to life. Inserting the coins into the slot machine of the Electricity Supply Board you muttered darkly about landlord fraud and one-armed bandits. A turn of the clockwork key brought light and warmth back into the bed-sit but you feared for the gastronomic wonder whose rise to perfection had been so brutally interrupted by the amusement-arcade fickleness of the coin slide.

When it was cold enough, and your breath was a fine mist, you stood around the schoolyard with the other worldly souls smoking imaginary cigarettes. Patting your pursed lips with forefinger and second finger slightly parted, and then pulling them wearily away in one long exhalation, the smoke trails of cold made the effect more compelling. So the shivering ten-year-old spivs would hop from one foot to the other in Firhouse National School waiting for life to catch up with their aspirations.

North American and Continental visitors were always puzzled by the Iceland of Ireland. A nation of carnival folk living in igloos. Watching the mini-skirted, blue-legged Young Ones go screeching and tottering up O'Connell Street, Gary or Françoise would shake their bescarfed and behatted heads in shivering disbelief. You once met a French language assistant who spent a whole weekend reading in the hot press. Not the victim of some foul fiend or an unannounced pogrom but a stray abandoned by the heating system, turned off by her host family who had gone away for the weekend. There were, of course, those in love with the adventure of winter, with the bracing draught of duty. Outdoor types and social reformers, alarmed at the overheated inwardness of the idle and the buttery indifference of the pampered. They loved Christmas swims and January hikes and were forever walking over to closed windows to raise the latch and empty out the patiently hoarded heat of an afternoon (*You don't mind, do you? It's a bit stuffy in here*). When you went to their

houses, you often feigned a cold to explain an extra jumper or made a pretext of grumpy eccentricity by sitting hunched in an armchair in a great, dark overcoat, like a St Petersburg court clerk down on his luck. You feared the sly tackle of the cold, the anxious parsing of the symptoms as they came – the sneeze that repeated itself once too often, not the random assault of dust or pepper but advance notice of the arrival of that irritable wretch, the common cold. The base of the nose raw with irritation. The disintegrating tissues and all the world seen through a snuffling fog.

There is a coal fire in the bank in the heart of the city. A real one, with crumbling logs, and not some sound and light show invented by a devious boffin with a dust allergy. But cold like pain is a dark monopolist and it is the poor not the rich who are tracked by its unmercifulness. On the television, the satellites promise high pressure and sunny spells. But the weather maps of want cover our towns and cities with their cold fronts and zero temperatures. Here, the careful sacrifice of the briquette and the fitful heat of the allowance are poverty's hopscotch of survival. And so the fires burn, a Paschal promise, a bright reminder of the endless, fugitive search for warmth.